The fact is that Planet Earth is almost depleted. We are running out of food. Population growth and resource depletion are on a collision course, and we seem powerless to avert a cataclysmic clash.

The secular spokesmen are not mincing their words. "The end is near," they say.

What is the Christian responsibility towards resources already near exhaustion? What can we do about the specter of worldwide famine? How should we grapple with the increasing problems of survival?

Is there *anything* we can do?

What does God expect from us?

W. DAYTON ROBERTS

A compelling look at the current state of Planet Earth

RUNNING OUT

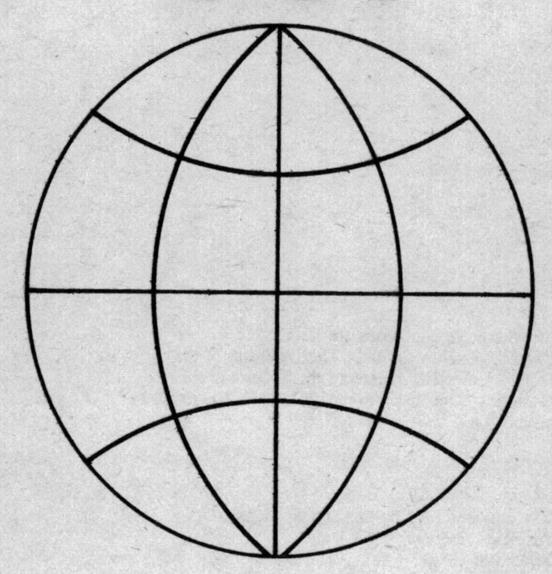

A Division of G/L Publications
Glendale, California, U.S.A.

The Scripture quotations are from *The New English Bible.* © The Delegates of the Oxford University Press and the Syndics of the Cambridge University Press 1961, 1970. Reprinted by permission.

Published by
Regal Books Division, G/L Publications
Glendale, California 91209

Library of Congress Catalog Card No. 75-5168
ISBN 0-8307-0366-7

CONTENTS

FOREWORD

Martin Niemoller was a German pastor imprisoned by the Nazis during World War II. It was in prison that he experienced what he called his "second conversion." Niemoller had so despised the atrocities of the Hitler regime that he came to hate the prison guard who brought him his food each day. Seeing the Nazi insignia on his uniform, all the indignation and outrage Niemoller felt toward that evil system was directed against that guard. Then, one day, Niemoller tells how he suddenly realized that Jesus Christ died on the cross, for that guard; that Christ loved him that much. And in the same fashion, Niemoller was bound to love that guard, and love every man. The atonement of Christ took on a whole new meaning for

Niemoller. Its implications were revolutionary, for *every* person was to be loved with the sacrificial love of Jesus Christ.

As Neimoller learned, the message of God's redeeming love in Jesus Christ is inseparable from the living witness of that love, in each of our lives, for others. There are two great commandments, which together as one make the whole gospel—to love God with all our heart, soul, strength and mind, and to love our neighbor as ourselves. "If you heed my commandments, you will dwell in my love," our Lord said (see John 15:10). If our lives have been truly touched by the miraculous, redeeming love of Christ, and if we are dwelling in that love, then we will be living forth the wholeness of the gospel. In all that we do and say our lives will be a natural witness of Christ's love for every person.

Related to the artificial separation we have made between the social and the spiritual is our neglect of the natural world in our theology and in our philosophy of life. We refer to the beauties of nature in our worship and thank God for these evidences of His power and infinite intelligence. Only recently, however, have individual Christians begun asking about how we are to relate to our environment.

The Scriptures teach us of the wholeness of creation. Man isn't alone in the universe. He is not independent from the world around him. He is unique in that he is part of creation but also is charged with caring for the creation. Too often we have interpreted the command to subdue the earth as permission to exploit it for our own ends. But we have forgotten that our ends are not always pure. The sin which separated us from our creator is the same sin which tarnishes our judgments and designs for the use of nature.

What is needed, then, is an understanding of the individual and corporate responsibility which man has for the welfare of his environment. Admittedly, it hasn't been until recent years that we have acknowledged that the resources of the earth are not automatically renewable.

For years we have been developing a society based upon the throw-away ethic, believing that somehow the trash of which we dispose today will find its way back into the production chain tomorrow. Or we have assumed that the resources required for the creation of that waste are limitless.

Russell Train, administrator of the Environmental Protection Agency says that the "throw-away attitude . . . includes built-in obsolescence, extravagant packaging, and throw-away products. But this attitude has to change. Much of the specific problems of energy shortages and inflation can be traced right to our growing wasteful habits. This waste is being encouraged by a product-oriented, advertising-stimulated economy."

Our concern for the preservation and even the enhancement of the world in which we live is not inconsistent with Scripture. In the words of the *King James Version* of the Bible we are to "replenish" the earth. The prophets and apostles of long ago, by the inspiration of the Holy Spirit, realized that God's people are responsible not only for a message of redemption relating to one's own sin, but for a message of redemption relating to the condition of the planet itself. This message must not only be one of word but one of action also. The people of God redeemed by

Jesus Christ are called to be "salt of the earth." While it is not exegetically correct to apply that phrase literally, I can't help but see a reference in it to the practicality of the Christian life. We cannot isolate ourselves from our world. And to do so is a sin of omission.

I am grateful to God for the growing insights which He is giving His people as new areas of need are discovered in our world. As I write this foreword I am preparing legislation to submit to the United States Senate which will prohibit the manufacture of aerosol products containing fluoro-carbons. A Federal study group has found these chemicals to be damaging upon the ozone surrounding the earth. According to the group, destruction of ozone can increase the incidence of skin cancer, caused by hazardous rays of the sun which ozone filters out.

If life is at all important we each have a responsibility to act upon facts which come to light relating to the care of our environment.

Doomsayers would have us believe that the end of the world is just around the corner. Perhaps we are near the end of the age. The Scriptures tell of a new heaven and a new earth. But while we remain, problems are opportunities for those committed to Jesus Christ to express His love toward all men in direct ministry to spiritual needs and in the redemption of a world in which all men live.

Mark O. Hatfield
United States Senator

ACKNOWLEDGMENTS

I wish to express my deep gratitude to:

Clayton L. "Mike" Berg, Jr., president of Latin America Mission Publications, for encouraging me to undertake this project even when it interfered with my normal responsibilities at LAMP;

My wife, Grace, who read and reread every belabored paragraph, patiently interacting with me and doing everything possible to keep me at the task;

Sociologist Jose Maria Blanch and biologist Rolando Mendoza, Christian ecologists and friends of long standing who have provided helpful stimulus together with John Strachan and Julio Garcia who are committed environmentalists of the next generation;

Many authors, some of whom I have quoted profusely and to whom I am indebted for facts and interpretations which I consider invaluable to an understanding of the ecological crisis, notably Drs. Barry Commoner, Paul Ehrlich, Francis Schaeffer, Samuel Moffett, Senator Mark Hatfield, the editors of *The Ecologist* (Edward Goldsmith, et al.), and the authors of *The Limits to Growth*;

Lois Thiessen, for reviewing and typing the manuscript.

AUTHOR'S INTRODUCTION

"The last days . . . ."

All sorts of Christian people have been issuing apocalyptic warnings from time immemorial. We have always believed in the imminence of Christ's Second Coming and the ensuing judgment although we haven't always lived like we believed it.

Now it's the secular scientists who are pronouncing doom.

"An examination of the relevant information available," warn the editors of *The Ecologist* of London, England, "has impressed upon us the extreme gravity of the global situation today. For, if current trends are allowed to persist, the breakdown of society and the irreversible disruption of the life-support systems on this planet, possibly by the end of the century, certainly within the lifetimes of our children, are inevitable."[1]

President Luis Echeverria of Mexico chooses his words carefully, but the apocalyptic note comes through clearly in his article, "Cooperation or Chaos":[2]

"The problems confronting man in the coming decades are so great that every effort must fail unless the overriding, categorical decision is taken to change and reorganize contemporary society .... Otherwise, we will have to accept as law the intolerable proposition that survival and development are to be the exclusive preserve of a small affluent portion of the world's population."

The fact is that Planet Earth is almost depleted. We are running out of food. We are running over with population.

"The major catastrophe will happen before the end of the century," predicted British statesman and author C.P. Snow as long as seven years ago. "We shall, in the rich countries, be surrounded by a sea of famine, involving hundreds of millions of human beings."[3]

And yet it seems unlikely that the rich will ever sacrifice voluntarily enough to feed the poor, especially when the latter are multiplying so much faster. Few scientists are prepared to be optimistic. The fact is, there isn't much to be optimistic about.

The secular spokesmen are not mincing their words. "The end is near," they say.

So where does the Christian stand?

St. Peter, inspired by the Holy Spirit, wrote 1900 years ago about the last days of Planet Earth. He posed a question:

"Since all these things will be destroyed in this way," he asked, "what kind of people should you be?"

It's a good question. What is the Christian responsibility towards resources already near exhaustion? What can we do about the specter of world-wide famine? How should we grapple with the increasing problems of survival? Is there *anything* we can do? What does God expect from us?

The following chapters attempt to offer a modest Primer of Christian Ecology. There is a rundown of man's abuse of his environment and some practical suggestions for coping with the death throes of Planet Earth. The information will help us to understand God's intent in creation, man's responsibility to nature, God's plans for planetary renewal, and remind us of the priorities for Christians in the "last days."

We have to change our attitudes. We've got to revamp our life-style. Each one of us must help avert disaster. We must share good fortune. We must help the starving brother. Now. Immediately!

Christ's kingdom must be announced. Quickly!

I wish I could write faster!

I wish I could express myself more graphically!

I wish I could get across to you the burden of growing concern that has enveloped me as I gathered the material for this book.

Time is running out!

The earth's biosphere is severely threatened.

Man is at the brink of disaster and needs to be warned of what is ahead of him.

Footnotes

1. Edward Goldsmith et al., *Blueprint for Survival* (Boston: Houghton Mifflin Co., 1972), p. vii.
2. Luis Echeverria Alvarez, "Cooperation or Chaos" in CERES (38), FAO Review on Development, March/April, 1974, p. 29.
3. Newspaper clipping, *New York Times*, November 13, 1968.

1

GOD'S ECOSPHERE

"The earth is the Lord's and all that is in it" (Ps. 24:1).

Man is a temporary tenant, a transient overlord. The earth is God's by creation, possession and purpose. It is His ecosphere, made for man's benefit and enjoyment and loaned to him in trust for the enactment of human history.

"The environment makes up a huge, enormously complex living machine," writes Dr. Barry Commoner, "that forms a thin dynamic layer on the earth's surface, and every human activity depends on the integrity and the proper functioning of this machine. . . . This machine is our biological capital, the basic apparatus on which our total productivity depends. If we destroy it, our most advanced technology will become useless and any economic and political system that depends on it will flounder. The environmental crisis is a signal of this approaching catastrophe."[1]

Ecological catastrophe was not always imminent. After each phase of the creative process in Genesis, God pronounced His handwork "good." The language of the story is simple and vivid. It evokes the wonder and richness of creation from formlessness to teeming life.

"But it is more than poetic," observe David and Pat Alexander in *Eerdman's Handbook to the Bible.*[2] "It tells us about what we need to know in order to understand ourselves and the world around us:

"The origin of the world and of life was no accident. There is a Creator: God.

"God made everything there is.

"All that God made was good.

"The high point of all God's creative acts was the making of man.

"Mankind is distinguished from all other creatures in two respects: he alone is made in God's own likeness; and he is given charge over all the rest.

"God's six 'days' of creative activity, followed by a 'day' of rest, sets the pattern for man's working life."

Building Blocks

The basic building blocks of life and matter are relatively few. Hydrogen, oxygen, carbon, nitrogen and less frequently elements like sulphur, phosphorus and various metals make up the environment. Each specific organic compound has its own proportions and spatial arrangements, with a resulting variety and complexity that is staggering. Solar heat is the motor for this unique creation.

Every living thing is dependent on many others, either indirectly through the physical and chemical features of the environment or directly for food and shelter. Scientists like to specialize and to focus on the minutiae of biology, chemistry and physics, studying how one particle bounces off another or one molecule reacts with another. The tendency is always to subdivide and analyze the detail. But this way we may miss the cybernetic relationships.

For example, tracing an element like nitrogen through the various environmental processes helps us to understand the ecosphere's complex unity. This is because the four chemical elements, carbon, hydrogen, oxygen and nitrogen, that make up the bulk of living matter, move in great interrelated cycles. Each element may at one time be a component of the air, now a part of some living organism, now forming a waste product in water or soil, and after a time perhaps built into mineral deposits or fossil remains.

Nitrogen enters the vastly complex ecosystem of the soil by nitrogen fixation or from the decay of plant matter and animal wastes. It is released slowly by the humus in the form of nitrates, which in turn are taken up by the roots of plants to become food for animals whose waste is then returned to the soil. If we could trace this cycle visibly, we could begin to understand how delicate is the process.

Energy is needed. The soil must be porous so that oxygen from the air and warmth from the sun can reach the roots where they oxidize the nitrogen. This porosity depends upon the humus, which also contains the major store of nitrogen, doubly enhancing its importance in the soil ecosystem. The process of life in the soil is not simply a chemical reaction, but a biophysical one as well.

The hydrosphere, or nature's water system, is host to a similar nitrogen cycle, except that there is no large reserve or organic nitrogen. In the aquatic ecosystems fish produce organic waste. Decaying microorganisms release nitrogen from organic forms and combine it with oxygen to form nitrate which is reconverted to organic forms by algae. The algae release important quantities of oxygen to the atmosphere and the algal organic matter nourishes small aquatic animals. They in turn are eaten by the fish, and the cycle is complete.

The air or atmospheric ecosystem is larger than that of the soil or of water and is the most uniform around the globe. Air is about 80 percent nitrogen gas, nearly 20 percent oxygen gas, with a very low concentration of carbon dioxide and much smaller quantities of rare gases such as helium, neon and argon, plus variable amounts of water vapor. The air is little affected by biological processes, except as it becomes polluted by heavy concentrations of foreign matter. The weather cycles keep the air in circulation and purify it. Anything that becomes airborne eventually is returned to the earth for reintegration in either the soil or aquatic ecosystems.

Ozone plays a particularly important role in the atmosphere. It settles in the stratosphere where it screens out the ultraviolet rays of the sun. This protects the sensitive living cells in the atmosphere and on the earth and the chemical reactions between them.

"In broad outline," summarizes Commoner,

"these are the environmental cycles which govern the behavior of the three great global systems: the air, the water and the soil. Within each of them live many thousands of different species of living things. Each species is suited to its particular environmental niche, and each, through its life processes, affects the physical and chemical properties of its immediate environment."[3]

Created Together

The Garden of Eden, as described in the first chapters of the Bible, was man's first ecosystem. "When the Lord God made earth and heaven, there was neither shrub nor plant growing wild upon the earth, because the Lord God had sent no rain on the earth; nor was there any man to till the ground" (Gen. 2:5).

First God caused the moisture in the earth's bodies of water to vaporize and then to precipitate in gentle rain over the earth's surface: "A flood (or mist) used to rise out of the earth and water all the surface of the ground" (v. 6). This was the normal cycle of the hydrosphere, activated by the sun's rays.

"Then the Lord God formed a man from the dust of the ground," that is, from the same elements which composed the ground. "And breathed into his nostrils the breath of life. Thus the man became a living creature" (v. 7).

"Then the Lord God planted a garden in Eden away to the east, and there he put the man whom he had formed. The Lord God made trees spring from the ground" (vv. 8,9). God was careful to ensure a supply of oxygen for the support of man's life-system. He planted trees "pleasant to look at and good for food" (v. 9). The warming rays of the sun, the refreshing shade of the trees, the photosynthesis of the leaves to manufacture oxygen, the falling and decaying of the leaves to enrich the soil's humus, were all part of the system.

Next were established the zoological ecosystems which are so dependent upon and so essential to those of the physical and botanical world. "So God formed out of the ground all the wild animals and all the birds of heaven. He brought them to the man to see what he would call them, and whatever the man called each living creature, that was its name" (v. 19).

Most important among these were the herbivores which fed off the greenery, thus preventing the suffocation of overgrowth, and restoring the content and structure of the soil's humus by contributing their wastes. It seems probable that after man's fall into sin and rebellion, some of these animals became carnivores and predators.

Woman

Fittingly, God's crowning act of creation was the woman, most beautiful and most ingeniously designed of all the creatures. " . . . for the man himself no partner had yet been found. And so the Lord God put the man into a trance, and while he slept, he took one of his ribs and closed the flesh over the place. The Lord God then built up the rib, which he had taken out of the man, into a woman" (vv. 20,21). God had "formed" the man from the elements of the earth. Now He took a sturdy piece of that man and "built up" the woman, presenting her to her husband as an equal but different creature.

The other animals had passed in review before the man to be named, one by one. The man had been pleased with them. But none of them offered him companionship. The woman, too, was brought before him, and when the man first saw the woman he fell in love with her on sight and burst into poetry: "Now this, at last—bone from my bones, flesh from my flesh!—this shall be called woman, for from man was this taken" (v. 23).

To one who loves the beauty of this earth, it is hard to imagine how God might have improved upon His creation. Repeatedly, God "saw that it was good." It was complete, self-sustaining and in balance. God Himself delighted in the beauty and productivity of nature. He specifically stated that He "cared for" the land. His eye was upon it in constant, loving appreciation.

Thus the creation bespoke God's tender attention, His all-embracing provision for the needs of mankind. And all nature, with its interlocking systems—from the orbits of the stars to the habits of the ants—became for the Bible authors a testimony to God's existence and to His concern for His creatures.

"When I look up at thy heavens, the work of thy fingers, the moon and stars set in their place by thee, what is man that thou shouldst remember him, mortal man that thou shouldst care for him?" (Ps. 8:3,4).

We can examine God's creation through the microscope that reveals, for example, the marvelous intricacy and tonal capacity of the inner ear. Or we can see it through the woodsman's binoculars and discover the upstream spawning habits of the salmon. Or again we can photograph it with the wide-angle lens of the ecologist and perceive the astonishingly complicated and interdependent order of the environment.

No matter what our perspective, we are forced to echo God's own evaluation after each stage of creation: "It is good!"

Footnotes

1. Barry Commoner, *The Closing Circle* (New York: Bantam, 1971-1974), p. 13.
2. David and Pat Alexander, *Eerdman's Handbook to the Bible* (Berkhamstead, England: Lion Publishing, 1973), p. 127.
3. Barry Commoner, op. cit., p. 28.

2

SIN AND STARVATION

Today's newspaper tells the pathetic story of 25,000 Hondurans isolated by mud, floods and tropical rains. They are starving to death just a few hundred miles from where I sit.

These people, rendered homeless by Hurricane Fifi a month ago, have seen their children buried in mud, and watched loved ones slip off into the oblivion of the dark brown water which swirled under the eaves of their tin-roofed shacks. They have fought over scraps of food dropped from planes. Now they are lethargic. They are just waiting to die.

The hurricane itself was not so devastating as the awesome landslides which slid into the valleys, damming temporarily the streams and rivers. The mud yielded eventually to the constant pounding of the water and together they rushed down upon the villages in an all-engulfing morass of liquid mud that felled trees and crushed houses, offering their inmates little hope of survival.

The 25,000 people remaining in the Aguan Valley of Honduras have been through that fearsome nightmare. They have not yet recovered, but they have survived. Most of the survivors are adults. The children didn't have the strength to struggle up through the tangle of mud, boards, and rafters until they could find something solid to hang on to.

And now the survivors are starving. Continuing rain has left them isolated, surrounded by miles of muddy lagoons with the roads gone and bridges washed out and landing strips buried. Only helicopters can bring relief, and they are too few and too busy elsewhere.

How do you explain these things? What is the cause of such tragedies? One is tempted to ask, just as Jesus' disciples once asked Him about a man blind from birth, "Lord, did this man sin, or his parents, that he should have been born blind?"

The Hondurans are dying because of sin, but not because *they* sinned. Like those killed by the falling tower of Siloam, or the Galileans slaughtered in the temple courtyard (Luke 13:1-5), they are the innocent victims of disaster. The disaster is a result of sin, but not specifically of the sins of its victims.

Hurricane Fifi, like the destruction of hurricanes, tornados, typhoons, storms, earthquakes, and volcanic eruptions, was an immediate result of the entrance of sin into God's perfect creation through the rebellion against God which Adam introduced and symbolizes. "Because you have listened to your wife and have eaten from the tree which I forbade you," God said, "accursed shall be the ground on your account" (Gen. 3:17). On account of man and his sin, Planet Earth was cursed with upheavals—floods and drought, storms and earthquakes—which have made man's life painful and difficult. Some of these phenomena may have existed before man sinned. The destructive character is a part of the "curse" following the "fall."

"For sin pays a wage," Paul says, "and the wage is death."

We usually think of this as applying to man himself, the sinner. We know that man is spiritually dead—separated from God. We also know that he is morally dead—his attitudes warped by sin, his motives twisted, his values distorted. Man cannot of himself live as God would want him to live. His life is dominated by selfish desires; his relationships soured by egocentric and inimical acts. His thoughts, words and deeds are permeated with the sickness of sin and corruption.

Physically, man is in the process of dying. The final state of physical death is only in process of becoming a reality. Illness, weakness, rot and corruption are a foretaste of what lies ahead of every man.

Earth's Dying

The death wage of sin is not limited to us as individuals, nor to our families, nor to the human race. "Accursed shall be the *ground* on your account." The biosphere has reaped some of the consequences of man's evil bent. It has fallen heir to the natural catastrophes which characterize the environment of fallen men, and it has been forced to endure the havoc that sin-tainted stewards of nature have wrought upon her.

God created man as the crowning act of His cosmic power. And He put him as ruler over the rest of creation. Man became God's steward, His vice-president for earthly affairs, His minister of environment. Man has failed dismally, tragically, catastrophically.

Instead of an orderly world, with a balanced ecology and adequate resources equitably distributed, we have an earth abused and exploited. Its mountains are denuded of forest, its hills and valleys eroded and sterilized, its atmosphere laden with toxic and obnoxious gases, its waters polluted and contaminated, its minerals depleted, its animal life eliminated. The world is characterized by malnutrition, hunger, pestilence and death.

Man's stewardship has been characterized by negligence, shortsightedness, exploitation, abuse and outright rape.

The devastation in Honduras is evidence of this. Careless clearing of the hillside forests caused erosion and made possible the deadly landslides which dammed the rivers and then burst upon the populated valleys with an almost diabolical synchronization. Excessive use of herbicides and chemical fertilizers may have aggravated the situation by debilitating the structure of the soil, making it more vulnerable to erosion and slippage. Hurricane Fifi was the fuse, not the explosive, that "blew up" the northern coast of Honduras. The real culprit was ecological ignorance.

"The environmental crisis tells us that there is something seriously wrong with the way in which human beings have occupied their habitat, the earth," concludes Dr. Commoner. "The fault must lie not with nature, but with man. For no one has argued, to my knowledge, that the recent advent of pollutants on the earth is the result of some natural change independent of man. Indeed, the few remaining areas of the world that are relatively untouched by the powerful hand of man are, to that degree, free of smog, foul water and deteriorating soil. Environmental deterioration must be due to some fault in the human activities on the earth."[1]

As an ecologist, Commoner describes it in humanistic terms. But he is talking about nothing more or less than the sin of irresponsible management. Man is an unfaithful steward. He has violated his trust. He has failed his God, damaged his environment and condemned himself to destruction.

Man's abuse of his environment is nothing new. "Few areas of the earth have borne more human impact, for good and ill," says James Houston, "than the lands of the Bible."[2] Here many animals were domesticated, irrigation systems initiated and towns created. "But here man has also destroyed the vegetation cover, induced soil erosion, and possibly even climatic deterioration."

A Rich Land

God's own love for nature was the basis of His charge to the Israelites as they entered the Promised Land of Canaan. "The land which you are entering to occupy is not like the land of Egypt from which you have come," He explained to the Hebrews, "where, after sowing your seed, you irrigated it by foot like a vegetable garden. But the land into which you are crossing to occupy is a land of mountains and valleys watered by the rain of heaven. It is a land which the

Lord your God tends and on which his eye rests from year's end to year's end" (Deut. 11:10-12).

"Thus the Hebrews had no word for nature," continues Houston, "other than the idea of the activity of God himself. It was God who spoke in the thunderstorm. He blessed in the rainfall; he cursed in the drought. God breathed in the wind, as he judged in the earthquake and manifested his glory in the heavens."

The Mediterranean climate, its flora, fauna and soils—its ecosystems—are delicately balanced. Since war could jeopardize this equilibrium, God cautioned the Israelites as to how they should possess the territory. "I will not drive them out all in one year, or the land would become waste and the wild beasts too many for you. I will drive them out little by little until your numbers have grown enough to take possession of the whole country" (Exod. 23:29,30).

Again, the children of Israel were reminded of the ecological and economic value of trees. "When you are at war, and lay siege to a city for a long time in order to take it, do not destroy its trees by taking the axe to them, for they provide you with food; you shall not cut them down" (Deut. 20:19).

In one instance God instructed the Israelites to liquidate the threat of a Moabite rebellion by practicing the common techniques of contemporary warfare. The Hebrews "entered the land of Moab, destroying as they went. They razed the cities to the ground; they littered every good piece of land with stones, each man casting one stone on to it; they stopped up every spring of water; they cut down all their fine trees" (2 Kings 3:24,25).

It is easy to understand why Israel, once a land

flowing with milk and honey, today is so arid, rocky and barren. Its modern settlers are struggling to recreate humus on the stony terraces, to reforest the barren hills and to restore water systems to the desolate plains.

It is not our purpose here to document man's ravaging of nature. We want to establish the link between man's irresponsibility and ecological ruin. As the one creature of God who is capable of controlling the ecosystems, man is the doubly-responsible "minister of environmental management" and is accountable for the disasters we face.

A Diet of Death

My generation grew up on a diet of war pictures. They formed the backdrop of the last 40 years of human experience. We carry a mental kaleidoscope of gruesome impressions—bodies strewn on the beaches of Normandy and Iwo Jima, mushroom clouds over Hiroshima and Nagasaki, gaunt corpses in the death vans of Auschwitz, half-burnt and crippled children in the rice paddies of Viet Nam.

The symbol of death for the coming generation is not going to be war but starvation, what Richard Selzer has called "strangulation in the open air."[3] And the pictures that will be most vividly remembered during the next thirty years will be those of swollen bellies and matchstick arms, flat pendulant breasts and emaciated faces, carcasses of cattle scattered on parched and cracking deserts. It will be as close as our TV screens.

Death by starvation is a unique process. The starving person in a very real sense feeds upon his own body. In its frantic search for glucose to fuel the brain and sustain the body's energy, the liver first exhausts its own glucose reserves. Then it turns to the tissues of the body and breaks down into glucose the protein of the muscles and other body organs. "Fat is seized from its various depots," Selzer goes on to explain, "and is changed to fatty acids in the liver, which substances supply energy to the remainder of the body. So it is that like the praying mantis, we turn upon our own bodies to feed, eating ourselves with the mindless voracity of insects biting off their own legs."

The body resorts to many tricks for survival. The starving person is less active, requiring fewer calories than normally. The body learns to substitute ketones for protein as brain food, thus retarding the disintegration of muscles to the point where the victim would no longer be able to move about in search of food.

"The child," continues Selzer, "entering a period of starvation, stops growing, for growth is a luxury, demanding reckless amounts of energy. Nor is this growth retrievable . . . . If the deprivation takes place during the first year of life, the effect is even more terrible, for during that first year the brain is still in a stage of development. Without energy to grow, the brain is dwarfed, experiences fewer cell divisions, forms less mass: there is stunting of the intellect as well.

"It is easy to see that the underfed pregnant woman carries a baby that is highly likely to have noticeably hampered mental faculties. The significance to the world of this fact is immeasurable. We have not yet reaped the whirlwind of the famines that have pinched the brains of millions of people on several of our continents in the past ten years. Consider, if you can, the effect on society of millions of adults who have undergone such periods of starvation in infancy, and who bear the scars of their cerebral devastation."

Thus it is, and thus it will be, in ever-increasing proportions, that man reaps in starvation the harvest of his abuse of the environment. The sinfulness of man and the devastation of his environment are parts of the same package.

The starvation of millions of people during the years just ahead will bring this home to us as nothing else could!

Footnotes

1. Barry Commoner, *The Closing Circle* (New York: Bantam, 1971-1974), p. 122.
2. James Houston, "The Bible in Its Environment" in *Eerdman's Handbook to the Bible* (Berkhamstead, England: Lion Publishing, 1973), p. 10ff.
3. Richard Selzer, "Strangulation in the Open Air" in *Harpers* magazine, June, 1974, p. 16.

3

POPULATION AND THE LILY POND

Until Thomas Robert Malthus came along no one worried very much about overpopulating the earth. There was plenty of room. Space had never been a problem. And although raising food in hostile soil was a lot of work, there was always more arable land available.

John Maynard Keynes has called Malthus "the first of the Cambridge economists."[1] As a child he knew and admired Jean-Jacques Rousseau and David Hume, both of whom were friends of his father and frequent visitors in the Malthus home. When Malthus wrote his famous *First Essay on Population* he was building on concepts already expressed by Hume and by Adam Smith. This *Essay*, first drafted at the end of the eighteenth century and frequently updated thereafter, has become the classic in the science of demography. He was about thirty years of age when he wrote it.

Making two basic assumptions—that food is essential for man's existence, and that sexual activity is necessary and will remain more or less constant—Malthus asserted that "the capacity for population growth is infinitely greater than the capacity of the earth to produce food for mankind."

"Population," he continued, "if it encounters no obstacles, increases in geometric progression. Food can increase only in arithmetic progression. Only the most elemental knowledge of numbers is required," Malthus concluded, "to appreciate the immense difference in favor of the first of these two forces."

He calculated the population would double about every twenty-five years. In 200 years—at the end of the twentieth century—the capacity of food production would have increased ten-fold, and the population would have increased 512 times!

"We have not assigned any limit to the production of the earth," Malthus stated. "We have conceived of it as being capable of indefinite increase." He went on to reason that even so, man would not be able to produce enough food, and this would put cruel but inexorable limits on the growth of population. Unchecked demographic expansion would ultimately be cancelled out by massive and grotesque starvation.

He has proven to be a good prophet. His conclusions have turned out to be correct, although the timing was not precisely as calculated. Population growth has reached a frightening high in recent years. Each day the sun sets on 200,000 more people than were there to witness the sunrise. Two—4—8—16—32—64—128—256—each time the rhythm of exponential growth moves faster until, like Ravel's *Bolero*, the tempo of the dance becomes wild, insane, uncontrollable.

Looking Ahead

The basic question is not one of space on the planet, but of food. Can the resources and life-systems of the earth support and sustain so many people even if the distribution of food is equitable?

There are those who answer with an outright "No!" They predict an increasing number of local famines, like those in the Sahel, Ethiopia, central India and Bangladesh, which eventually will turn into world-wide hunger, with the poor nations struggling to get a larger share of the earth's resources and the rich nations fighting to maintain their standard of living.

Others are more optimistic, albeit guardedly so. Their optimism is predicated on voluntary cooperation, a cut-back on unbridled technocracy, international controls and altruistic attitudes on the part of the wealthy nations towards the less fortunate. They admit that a drought or any other natural calamity could upset their precarious calculations.

The Club of Rome came into the limelight a few years ago with a very pessimistic set of predictions. An informal group of eighty-five leading international businessmen, scientists and thinkers met for the first time in Rome (hence the name) to devote themselves to developing ways of dealing with an ever more complicated world. They computerized their findings and published the results in a now famous book, *The Limits to Growth.*[2]

This book sustains, with valid facts and reasons, that exponential economic and population growth cannot continue indefinitely on this finite planet. It predicted that unless the explosion is halted, civilization will grind to a standstill, and mankind will either starve or suffocate in pollution. A time factor of less than a century was suggested.

These findings were challenged by some sociologists and agroeconomists, and as a matter of fact a mathematical error was discovered in the computer model on which the predictions were based. A recent updating of the Club of Rome's appraisal is not much more encouraging.[3]

The new study hammers home the urgency of action required to avert global disaster. Only by tackling immediately every aspect of the problem can we avoid turning present "catastrophic" food shortages in India and Africa into what is described as "apocalyptic" famine by the year 2010. It does not insist on the bleak alternative of no economic growth, but favors a process of selective expansion. A truly interwoven global economic system is needed in which all nations help one another for the benefit of all.

More than Evangelism

In the lobby of the great Kongresshalle in West Berlin, the sponsors of the first World Congress on Evangelism in 1967 set up a huge population clock. Throughout the workshops and assemblies, the lunch hours and the coffee breaks—even during the night when the congress hall was closed to delegates—the digital clock clattered on inexorably. Computing the addition of more than two new inhabitants on the earth per second, or 150 per minute,* the clock was an awesome reminder of the urgent task of Christian witness facing the delegates.

It was significant that almost without exception the public references to the population clock during the congress indicated that it was viewed only as a challenge to evangelism. "Every day there are more people who need to hear the gospel," the clock was made to say. "Work, therefore, while you can, to get the Good News out."

This is a legitimate and significant line of reasoning. There seemed to be little awareness of any possible relationship between the "population bomb" and the Apocalypse. The second coming of Christ was no more "imminent" than it ever had been. Nor was the demographic explosion seen as a threat to man's very survival on the earth.

My own awareness of the acuteness and urgency of the demographic problem came a year earlier, on a visit to India in 1966. Our flight from Bangkok landed in Calcutta in the late evening. There we were met with a note from our missionary host stating that the people of Calcutta were rioting in the streets in protest against the scarcity and high cost of food. Therefore it would be wise, he said, not to try to come into town at all, but to reboard the same plane and continue to Bombay. This we did.

Arrival in Bombay was after midnight. As we rode the bus from the airport into the city, my first surprise was to discover that not everybody was asleep. We

*N.B. In 1974 demographers calculate the population increase per second as slightly under two, or at most, 120 per minute.

passed the university campus. There lights were blazing and classes were in session at 1:30 A.M.! We soon learned that Bombay lived in two shifts—day and night. Apparently there were more people than beds. There were people sleeping in the railroad station, in the doorways of houses, in the patios and gardens. Some slept by night, others by day. And in the restaurants a small notice on the menus warned that "Only one entrée per customer may be ordered."

In Bombay, at least, overpopulation had already arrived.

The problem is not limited to India. No sector of society, no nation, no part of the globe can remain unaffected by the human race's exponential growth rate. According to the most authoritative calculations, taking into account present trends and anticipating measures to control the demographic upsurge, in the next ten decades Europe will increase in population by 50 percent. Populations of North America, the Soviet Union and the Far East will double. During the same period Oceania will increase 2.5 times, southern Asia 4 times, Latin America 5 times, and Africa almost 6 times over![4]

Dr. Paul Ehrlich of Stanford University and author of *The Population Bomb*, was asked in an interview, "Why do you say the death of the world is imminent?" The distinguished population biologist replied, "Because the human population of the planet is about five times too large, and we're managing to support all these people—at today's level of misery—only by spending our capital, burning our fossil fuels, dispersing our mineral resources and turning our fresh water into salt water." Dr. Ehrlich went on to point out: "We're adding 70 million people to the world each year. This means we have a new United States—in population and all that implies in environmental stresses—every three years."[5]

Controlling Death

The reasons generally given for the unparalleled surge in population have to do with the advance of medical technology. Death control has outpaced birth control. Man's life-span is being lengthened by modern miracle drugs. At the same time infant mortality is dropping dramatically.

"The single thrust that did most to induce the population crisis," according to sociologist Barbara Ward, "was the cure of malaria and other epidemics—virtually a by-product of the Second World War. It meant that into societies which were unmodernized in almost every other way, we suddenly put in effective modern health control."[6]

When the Western World went through its own process of development, Ms. Ward points out, diseases like malaria, cholera, and typhoid were still prevalent. In fact, they were not controlled until long after the industrial apparatus was in place.

"We were absolutely unprepared for what would happen if public health were introduced ahead of any other forms of development such as education, changes in the role of woman, movement out of agriculture, urbanization, industrialization. All these changes," she explains, "which were either concomitant with the change in health in the Western World or actually ahead of it, have lagged in the developing world while health has been effectively modernized. As a result, the death rate could be as much as halved in a few years—as in Ceylon—and since birth rates were maintained, population doubled in only twenty years."

This would seem to support the thesis of many demographers, among them notably Argentines, Brazilians and Mexicans, that social change is more important than birth control to reduce population growth. Over the long haul, this assumption is probably correct, inasmuch as demographic growth has historically tended to level off in the developed countries with the arrival of affluence.

A few authorities, notably the late Josue de Castro, brilliant Brazilian sociologist and author of the classic *Black Book of Hunger*, try even to make hunger the *cause* rather than the *effect* of overpopulation. De Castro claims that a protein-rich diet lowers the fertility index, while a diet deficient in animal proteins produces higher fertility. This is why the poor nations multiply so fast, while the population of the wealthy nations is stabilized.[7]

Barbara Ward's explanation for the lower birth rate of developed countries is much more credible: ". . . education, opportunity, status of woman and, above all, the slow but steady increase in opportunity, hope and justice have been infinitely more important," she says. Even the techniques of family planning have played a secondary role in the leveling-off process. "The introduction of more general and varied kinds of family planning in the twentieth century has only confirmed trends which were already evident. The history of the nineteenth century suggests that profound social change is fully as important—and possibly more important."[8]

Ms. Ward demonstrates that the decision-making process involved in intelligent family planning cannot be carried on in utter ignorance of biology, economic possibilities and society's needs. "I fear," she says, "educated people have a tendency to underestimate the degree to which ignorance is a blanket on all responsible decision-making . . . . It is quite clear," she adds, "that part of the tragedy in all societies in which women are subject creatures is that those who carry the chief physical burdens have no say.

"This is critically important. One of the key factors in the Western World, in Russia and probably China is that women are more involved in the processes by which decisions are taken. Any society dedicated to 'machismo' will continue to have a high birth rate."

There is no place in such a society for responsibility. She cites as an illustration the number of abandoned children in Latin America. "There are up to a million children between the ages of two and ten who live the life of vermin and grow up beyond any reach of human development or dignity."

Justice and Population

Education and a more dignified and human status for women are part of a wider issue. "No policy for responsible parenthood by whatever means is possible where malnutrition, illiteracy, shack-dwelling, unemployment and sheer dark despair are the family's daily lot . . . . The division of the world into careless rich and despairing poor is the chief reason why there will be no effective, acceptable population policy in the world unless there is an acceptable and accepted policy of social justice . . . .

"New facts and statistics have been appearing in recent years," concludes Ms. Ward, "and they suggest very strongly that the achievement of local social justice is the quickest route to a greater stabilization of family size . . . . The failure of aid to reach the mass of the people has been the weakness of development policy over the last twenty years . . . . And the places where population is beginning to stabilize are those in which social justice has been thrust down to the mass of the people."[9]

Which comes first, the chicken or the egg? Another Christian sociologist, Dr. Jose Maria Blanch, inverts the order but comes up with essentially the same answers.

"Assuming a nuclear holocaust can be avoided," he affirms, "the transcendental problems of our time are: first, to achieve a world in which human beings control their fertility to the extent that the population does not exceed a limit in which the resources are sufficient for all, and secondly, to create the social institutions and common human values that will make this world a more acceptable place to live."

Dr. Blanch bases his reasoning on the Christian concepts of responsibility to man's Creator and love to man's neighbor. "In Jesus Christ we are told that this love is the meaning of life and its foundation," he states. "The fact that no 'Christian' social system exists—nor any just one—should not keep us from seeking a system that is better and more just," he concludes. "What we can do is to reflect critically upon the possibilities that arise in the course of our efforts. It is the duty of every Christian and of the Church to revive and promote the concept of the responsibility of all for all, the idea of concrete responsibility in mutual unity."[10]

There is an important ecological principle underlying the insights of both of these sociologists: namely, man's social and biological systems are interrelated and interdependent, as are all the ecosystems of our planet. This simply means that the battle for survival has more than one front, and we cannot make the simplistic mistake of counting on birth control alone to reduce human overpopulation. Social liberation and social justice are at least as important. Attitudes are more effective than IUD's. As Paul Ehrlich points out, it's not the *unwanted* babies who are causing our population problem so much as the *wanted* ones.

In any case, the problem is frighteningly acute, and Ehrlich makes no apologies for the apocalyptic character of his projections. "I *am* an alarmist," he admits, "because I'm very (expletive deleted) alarmed. I believe we're facing the *brink* because of population pressures. I'm certainly not exaggerating the staggering rate of population growth: it's right there in plain, round numbers. Whatever problems I'm diverting attention from will be academic if we don't face the population-environment crisis now."[11]

It is true that some other scientists are more sanguine about the situation than is Ehrlich. They blame factors other than population growth for the food and environmental crises in different parts of the globe. Francis P. Filice, for example, claims that the world population is not all that overwhelming. "We could put the entire world population in the state of Texas," he says, "and each man, woman and child could be allotted 2000 square feet (the average home ranges between 1400 and 1800 square feet), and the whole rest of the world would be empty."[12]

This kind of reasoning seems plausible and is superficially comforting. But it overlooks two factors—that of the burden presently being placed on the finite biosphere by human proliferation and the exponential character of population growth.

In the next chapters we shall be examining the strain which demographic explosion is putting on our biosphere, both its renewable and its nonrenewable resources. Here we would like to add a further word about exponential growth and the unexpectedness with which its consequences can overtake us.

"A French riddle for children illustrates another aspect of exponential growth—the apparent suddenness with which it approaches a fixed limit. Suppose you own a pond on which a water lily is growing. The lily plant doubles its size each day. If the lily were allowed to grow unchecked, it would completely cover the pond in thirty days, choking off the other forms of life in the water. For a long time the lily plant seems small, and so you decide not to worry about cutting it back until it covers half the pond. On what day will that be? On the twenty-ninth day, of course. You have one day to save your pond."[13]

Despite conflicting opinions about the speed of the population explosion and its future projections, all scientists are agreed that demographic growth is exponential and we cannot allow ourselves to become complacent about the future outlook. "Such complacency could be followed by drastic and irremediable consequences," declares Philip M. Hauser, population sociologist. "To continue on the assumption that much yet remains to be done to bring excessive population growth under control is certainly the course of wisdom."[14]

Malthus was right. Left alone, the water lily will

choke the pond. And today is at least the twenty-ninth day of the month!

Footnotes

1. John Maynard Keynes, *Robert Malthus: The First of the Cambridge Economists* (London: Macmillan, 1933-1951), translated and published as a prologue to Thomas Robert Malthus, *Primer Ensayo Sobre la Poblacion* (Madrid: Alianza Editorial, 1966).
2. Dennis L. Meadows et al., *The Limits to Growth, A Report for the Club of Rome's Project on the Predicament of Mankind* (New York: Universe Books, 1972).
3. Club of Rome, *Mankind at the Turning Point*, quoted by *Time* magazine, October 21, 1974.
4. "Perspectivas para el Porvenir" en *El Correo*, revista de UNESCO, Mayo, 1974, p. 16.
5. Paul Ehrlich interview, *Project Survival* by Geoffrey Norman. (Chicago: Playboy Press, 1971), pp. 69,74.
6. Barbara Ward, "Environment, Population and Development" in *WACC Journal*, Frankfurt/Main, Germany, 1/1974 (vol. XX), p. 8.
7. Josue de Castro, *El libro negro del hambre* ("The Black Book of Hunger," Buenos Aires: Editorial Universitario, 1964, original in Portuguese, 1960), pp. 32,33.
8. Barbara Ward, op. cit., p. 8.
9. *Ibid*, p. 12.
10. Jose Maria Blanch, "Ecologia humana y explosion demografica" en *Certeza* (No. 54) Abril/Mayo, 1974, p. 178.
11. Paul Ehrlich interview, op. cit., p. 77.
12. Francis P. Filice, "Population Growth and Population Control, Another View (1970)" in *Population: a Clash of Prophets*, Edward Pohlman, editor (New York: New American Library, 1973), p. 152.
13. Dennis L. Meadows et al., op. cit., p. 37.
14. Philip M. Hauser, Testimony before special subcommittee, U.S. Senate (1971) in *Population: a Clash of Prophets*, Edward Pohlman, editor (New York: New American Library, 1973), p. 85.

4

OUR DIMINISHING ENVIRONMENT— LAND

Sometime between 1980 and 1990 the demand for land will exceed the supply.

This flat statement is made by the ecologists who authored *Blueprint for Survival.*[1] The calculation is based on the premise that demand for land must increase with population growth and that the per person requirement is 0.5 hectares (about 1 acre) for agricultural purposes and 0.08 hectares for nonagricultural purposes. If yields of agricultural products should in the interim double their present levels, the effect would be to add only thirty years to the world's food supply.

The total land area of the earth is approximately 32.5 billion acres (13.15 billion hectares), of which only 3 billion acres (1.21 billion hectares) are cultivated at the present time, and an additional 5.4 billion acres (2.2 billion hectares) are used for grazing.[2]

Most of the world's land surface is covered with icecaps, permafrost, deserts, forests and urban and industrial areas. The amount of marginal land available for expanding agriculture is severely limited. Its cultivation would be far costlier than that of the areas presently being farmed. In the most heavily populated areas of the earth such as the Soviet Union, China and the rest of Asia and Europe, there is little marginal land remaining for development. And the U.S. has only 50 million acres (20 million hectares) of unfarmed land which might be made available for agriculture.

"While some expansion is possible," *Blueprint* concludes, "it is unlikely that the resources of capital and materials can be made available to produce more than minor increases in food production from these sources."[3] The cost per hectare of opening up new farm land runs from $215 to $5,275 and averages out at $1,150.[4]

Any agricultural expansion in the Near East and Africa would require tremendous investments for irrigation. The Amazonian jungle, sometimes called "the last green lung of the world" because of its vast rain forests, might provide some farm land, but the problems are immense. Experience in clearing the virgin forests of Central America would seem to indicate that the removal of primeval vegetation often triggers an erosion process leading to desert. The process is all but irreversible because once organic matter is exposed it tends to mineralize. The unstable soils of Amazonia are some seventy feet thick, but would probably erode very quickly if exposed to the equatorial climate. When Khrushchev cleared the forests of Kazakhstan for agriculture he left a dust bowl of some thirty million acres.

Measured against the high costs of small productivity of cultivating marginal lands is the fact that existing agricultural land will most likely be reduced as demands for urban and industrial development increase. Land will be required for housing, roads, airports, factories, and office buildings.

The limitations of arable land are pushing agronomy into intensified production. But beyond a certain point, which may vary according to climate and soil conditions, the intensification of farming causes soil deterioration and eventually erosion. This is the complex result of chemical fertilizers and pesticides, non-rotation of crops, use of heavy machinery and overstocking of farm animals. All of these speed the process of deterioration of soil structure, with attendant problems of drainage, waterlogging, salinity and lowering of water tables.

Erosion of farmlands in some areas is associated with the spread of deserts. In 1882, 9.4 percent of the earth's surface was desert or wasteland. By 1952 this had increased to 23.3 percent of the land surface.[5] If the processes of erosion and the development of desert-wasteland have continued at the same pace since 1952—and it seems probable that the tempo has actually accelerated with the increase in deforestation and intensive farming—the present area of desert and wasteland could be calculated at about thirty percent of the earth's land surface.

It may not be today, or even tomorrow. But very soon, certainly before the end of the century, we shall be running out of arable land. And in some places this has already happened.

But worse than that is the way we are treating the land which we presently have!

Forming Dry Lands

A classic tragedy is the fate of that vast sub-Sahara area of equatorial Africa known as the Sahel. Two and a half million square miles, 6.48 million square kilometers, one-fifth of the continent is slipping out of human use. The area stretches across parts of Mauritania, Senegal, Mali, Upper Volta, Niger, Nigeria, Chad, Sudan, Kenya and Ethiopia. Always parched for rain and niggardly in its life-sustaining capacity, the Sahel's thin blanket of coarse grass, thorny bushes and scanty trees is now giving way to sterile sand.

It is not true that the Sahara desert is advancing inexorably southward for reasons that are a mystery. The Sahara is indeed growing in some places up to twenty or thirty miles per year. But the new desert areas are not always contiguous to the Sahara, and in any case the reasons can be easily understood.

The normal rainfall of the Sahel has always been minimal. But in 1968 came one of its periodic droughts. The plight of the Sahel was well publicized during the following years. Immediate relief was required. Geologists discovered vast reservoirs of underground water, however, so the developed nations rushed to the aid of the drought victims and provided not only food for the interim ($175 million in 1973 or half a million tons of food), but also deep wells and waterholes for the renewal of the land.

The existence of more water has encouraged the

increase of livestock herds, the principal industry of the area. These in turn have overgrazed the meager pasture and trampled the areas adjacent to the waterholes into dust bowls. The large flocks of goats, which eat vegetation down to the roots, can be blamed for much of the devastation. The net result has been the wholesale death of cattle, not from thirst but from lack of food. And what was already one of the most destitute portions of the globe was reduced even further to absolute barrenness and mass starvation.[6]

In other parts of the world the same phenomenon is taking place on a lesser scale. Dr. Rolando Mendoza, a Costa Rican ecologist, has observed in Latin America the irreversible ill effects of overgrazing. In different parts of the southern continent deserts have appeared, due to the excessive number of goats.

In Central America it has been the irresponsible felling of forests which has created arid and unproductive zones. The frontiersman is usually a "slash and burn" farmer. He knocks down the forest. Sometimes but not always, he leaves a few of the better trees standing. He burns the fallen vegetation on the ground. Then he plants rice and is able to get one or two good crops out of the soil. It is then either planted in pasture or is allowed to grow wild for a couple of years after which it is again slashed and sometimes burned and this time planted in corn, for which the fallen secondary growth provides adequate nutrients.

Except for the rapid depletion of forests, this procedure is not altogether bad, although erosion is intensified. But Mendoza reports that in many instances the forest is destroyed in order to plant cotton, and after a few years the latter is abandoned to leave the land converted into semi-desert. "Erosion is the cancer of the earth," Mendoza says, "and man helps to propagate it."[7]

In Costa Rica, a relatively enlightened nation, 12 million trees were felled in 1973 and only one million were planted. The commercial value of lumber multiplies the long-range economic loss represented by these statistics. By felling the forests man is depriving himself of sources of oxygen, contributing to the destruction of the soil, exposing the land to erosion and the populace to floods and landslides, and is probably changing the climate to his own detriment.

Fortunately, forests represent a resource that is to a degree renewable. In many places a laudable effort is being made in this direction. On a global scale, however, mankind has long been overdrawn at the bank on his forestry account.

Drugging the Land

One final insult to the land requires mention here.

A remarkable development of modern agricultural science has been the cultivation of new varieties of grains which respond dramatically to the use of chemical fertilizers to produce huge increases in the per-acre yield of these basic crops. In the State of Illinois when very little nitrogen fertilizer was used, the average corn yield was about 50 bushels per acre. In 1958, when the use of fertilizer had been multiplied ten-fold, the yield was about 70 bushels. In 1965, with a further increase in the use of fertilizer to 4,000 percent of its 1945 level, the yield was pushed up to about 95 bushels per acre. Even though the rate of increase slowed in proportion to the amount of fertilizer used, the farmers could afford to fertilize massively because the manufacturing cost of fertilizer kept going down over that period.[8]

Overlooked were two factors.

First, an immediate threat to public health began to emerge in the high nitrate and nitrite content of nearby surface waters. Citizens of Decatur, Illinois, began to discover that the nitrogen pollution of their streams, rivers, lakes and water supply increased and decreased in the ratio of seasonal fertilizing programs. Almost certainly, the relationship was one of cause and effect. The matter became one of great community concern, since the nitrate levels exceeded those recommended by public health authorities.[9]

Nitrate itself appears to be relatively innocuous in the human body. However, by action of certain intestinal bacteria, it can be converted to nitrite, a toxic chemical that combines with hemoglobin in the bloodstream, converting it to methomoglobin and preventing the transport of oxygen by the blood. These particular intestinal bacteria are usually more active in infants than in adults. The classical clinical effect of nitrite poisoning is to make an infant turn blue, with the serious danger of asphyxiation and death.[10]

A second consequence of programs of massive fertilization with inorganic chemicals is that in a very real sense the farmland gets "hooked on drugs." To achieve each step-up in production per acre, a greater amount of fertilizer is required each time until the ceiling of that particular genetic strain of grain is reached. The per-acre production levels off and eventually begins to decline. But worse than that is the fact that with artificial nitrogenization the soil either desists or is hindered from carrying on its own organic process of fertilization. With the increase of inorganic fertilizer, the natural fertilization process is snuffed out, requiring still more inorganic chemicals to replace the loss, and so on.

Finally, as a result of their specialized breeding, the chemically fertilized varieties of grain are much more vulnerable to pests and insects. They are dependent upon the broad use of pesticides and insecticides to ensure high production levels.

The "victories" of the "green revolution" are thus to a degree illusory. With the recent jumps in the cost of fertilizers (derived from petroleum), there is a real question as to whether the starving nations of the Third World can afford to experience a "green revolution."

The only long-term answer seems to be organic farming.

During World War II, Great Britain found herself increasingly isolated from important overseas sources of food by the supremacy of the German submarines in the Atlantic. The British government began an intensive campaign for growing more food. Coupled with the exhortations to plant more and cultivate more were high pressure sales tactics to induce gardeners and farmers to use more fertilizers. The industrial and financial resources of the nation were marshalled to make available vast quantities of what was then called "artificial manure." This policy has endured, in ever increasing proportions, over the intervening years.

"One thing, however, was forgotten," declared Sir Albert Howard, the granddaddy of organic farming. "No satisfactory answer to the following question has been provided. What will be the final result of all this on the land itself, on the well-being of crops, livestock, and mankind?" Sir Albert contends that health—not only that of the plants themselves, but also of the animals and people who live off the products of the earth—begins in the soil.

"Will the grow-more-food policy have solved one problem—the prevention of starvation—by the creation of another—the enthronement of the Old Man of the Sea on farming itself? What sort of account will Mother Earth render for using up the last reserve of soil fertility and for neglecting her great law of return? Who is going to foot the bill?"[11]

As we run out of arable land on our terrestrial planet, these are the important questions to ask.

Footnotes

1. Edward Goldsmith et al., *Blueprint for Survival* (New York: New American Library, 1974), p. 106.
2. *Ibid*, p. 104.
3. *Ibid*, p. 105.
4. Dennis L. Meadows et al., *The Limits to Growth* (New York: Universe Books, 1972), p. 58.
5. Edward Goldsmith et al., op. cit., chart p. 105.
6. Claire Sterling, "The Making of the Sub-Saharan Wasteland" in *Atlantic Monthly* magazine, May, 1974, p. 98.
7. Rolando Mendoza, "La problematica ecologica" en *Certeza* (No. 54) Abril/Mayo, 1974, p. 182.
8. Barry Commoner, *The Closing Circle* (New York: Bantam, 1971-1974), p. 81 f.
9. *Ibid*, pp. 82,86.
10. *Ibid*, pp. 79,90.
11. Sir Albert Howard, *The Soil and Health* (New York: Schocken, 1947-1972), p. 76.

5

OUR DIMINISHING ENVIRONMENT—WATER

"Human activity has already contaminated all the world's oceans, the atmosphere and even the remote glacial icecaps of Greenland and the Antarctic. Most of the rivers are more or less contaminated, and many of them are nauseating open sewers . . . . The problem is . . . whether civilization can survive its own impact upon nature."[1]

Water is man's most essential and least respected resource. It has simply been taken for granted.

It was not until the seventeenth century that men began to understand the hydrographic ecosystem. Back in the 1500's, a French Huguenot by the name of Bernard Palissy learned to produce masterpieces of enameled ceramics which he called terracotta. This invention saved him from the Inquisition by which he had been condemned to die for his involvement in the activities of the Reformation. Because of his artistic potential, Queen Catherine de Medici rescued his life by appointing him to the court of Henry III.

Palissy was an amateur scientist in several fields, notably geology, and an astute observer of nature. In 1580 he published a book in which he stated that all springs and rivers were fed exclusively by rainwater. Until that time, and for another century thereafter, most scientists believed that only part of the surface water came from the rains. They thought some mysterious pressures of wind and tide pushed the ocean's moisture under the earth's surface and up into the mountains where the rivers were born.

It was not until 1668 that another Frenchman, Pierre Perrault, proved by actual measurement that over a three-year period the land surface drained by the River Seine received six times as much rain as was carried off in the waters of the river. Perrault was thus the father of modern scientific hydrology.[2]

Even today, the apparent abundance of water has fostered an indifference and unconcern which may all too soon lead us to an irreversible contamination of the hydrosphere, rendering it impotent to support human life on the planet. Man desperately needs water—in sufficient quantities, around the season, in the right places, and of an adequate quality, free of pollution and contamination—in order to live and to grow his food and to nourish his industries.

Like all of nature, the water system is delicately and intrinsically linked to the rest of our environment.

"Costa Rica has an enormous hydroelectric potential," says a newspaper report on reforestation, "but we are losing it by neglecting the hydrographic drainage beds and by our indiscriminate leveling of forests."[3] The trees of natural forests preserve the superficial humus, keep the soil aerated with their roots, allow the rains to penetrate deeply, and prevent erosion. They are essential to the regulation of surface water's quantity, quality and constancy. Thus the disastrous consequences of the hurricane-triggered floods in Honduras and the monsoon-triggered floods in the Philippines would certainly have been reduced by intelligent management of lands and forests.

Our most serious concern about the world's water today is the problem of contamination and pollution. A current newspaper report said that the waters in the popular Coronado and Ocean Park beach areas of San Juan, Puerto Rico, "fail to meet minimum pollution standards and should be closed to swimmers," according to a Federal and Environmental Protection Agency report.[4] Such findings are alarmingly common all around the globe.

"In these pioneer years of the ocean age," states Senator Claiborne Pell, "the damage done sometimes seems to exceed the benefit reaped. Beaches from England to Puerto Rico to California have been soaked in oily slime. Fish and wildlife have been destroyed. Insecticides, dispersed in the Rhine River, killed fish and revived fears of other lethal legacies that may emerge from our casual use of waterways as garbage dumps."[5]

Destroying the Water

Ignorance, indifference, or simple callous disregard of the consequences has brought us near to a point of no-return. On the one hand, the industrial take of the fishing industry apparently has reached its maximum level. And on the other hand Japanese fish-eaters are dying of mercury poisoning. Further contamination could be accompanied by cosmic risks both to health and to the proper function of the hydrosphere itself.

Until recently deterred, the U.S. Army was disposing of containers of chemical agents in the Atlantic. The future disposal of increasing amounts of radioactive atomic wastes is still an unresolved problem. Millions of acres of offshore seabed have been leased for drilling. Giant tankers around the globe are flushing their immense holds with seawater. And by uninhibited tradition, the hundreds of thousands of commercial and pleasure vessels that ply the world's waters are daily spilling their garbage, trash and sewage into the illusory vastness of the oceans. Thor Heyerdahl of Kon Tiki fame, crossing the southern expanses of the Atlantic in his reed boat, at times could not draw ocean water for brushing his teeth because of its pollution.

The problem is most acute in surface waters. The major sources of contamination are sewage, industrial wastes and agricultural drainage, the run-off of (*feed lot manure,*) inorganic fertilizers, pesticides and herbicides. Reference was made to agricultural pollutants in the previous chapter. We shall deal here only with the sewage problem, particularly as it is multiplied by the congestion of population in our cities.

"Eutrophy" is a term coined to describe what happens when an ecosystem is overloaded or "dies" because it has been forced to accept more waste than can be used to provide the materials for other processes. In such conditions the self-regulating mechanisms can no longer function properly and the waste simply accumulates. In other words, eutrophy has increased. If the process is not corrected the farmland so affected will turn to desert, the atmosphere will become gas, or the water will become a cesspool.

"Thus," says *Blueprint*, referring to the marine ecosystem, "if the cycle is overloaded with too much sewage, detergents, or artificial fertilizers that are nutrients to aquatic plant life, the amount of oxygen required to effect the decomposition of these substances by the appropriate bacteria may be so high that other organisms will be deprived of an adequate supply. If this goes on long enough the oxygen level will be reduced to zero. Without oxygen, the bacteria will die and a crucial phase in the cycle will have been interrupted, thereby bringing it rapidly to a halt. As a result, what was once an elaborate ecosystem, supporting countless forms of life in close interaction with each other, now becomes a random arrangement of waste matter."[6]

Lake Erie has an acute case of eutrophy. The careless use of what once seemed to be an inexhaustible waterway to the sea has rendered it unfit for human use of almost any sort.

Lake Erie is about 12,000 years old, a product of the great ice sheet that moved south to gouge out the beds of the Great Lakes and then melted to fill them with clear water. When it was first studied in the seventeenth century, it contained a large and varied population of fish. The waters were clear and sparsely populated with algae. The nutrient salts leaching into the lake were sufficient only for a small plant crop which was in turn limited by the animals which fed on the algae.

"Once algae grow in it, a body of water can sustain a complex web of life; the small animals that eat the algae; the fish that feed on them; the bacteria of decay that return organic animal wastes to their inorganic forms—carbon dioxide, nitrate, and phosphate—which can then support the growth of fresh algae. This makes up the basic, freshwater ecological cycle."[7] Lake Erie continued in this state of biological balance for another 200 years.

At the turn of the century there were changes in the fish life of the lake. The lake sturgeon previously harvested at the rate of one million pounds a year almost disappeared. By 1964 only 4,000 pounds were taken. The same thing happened with the northern pike. Early studies revealed chemical and biological changes in the lake which might be related to the loss of fish. At first there was no cause-and-effect relationship established. Pollution was not immediately apparent.

What had happened was that the oxygen of the lake was depleted. The oxygen level is "a very sensitive index of pollution. When raw, untreated sewage enters a lake, it contributes a good deal of organic matter to the water. The more organic matter entering the lake, the greater the amount of oxygen needed to convert it to inorganic salts: this 'biological oxygen demand' or BOD, is therefore a measure of the amount of organic pollution in the water."[8]

The mystery persisted because the amounts of sewage were not considered to be enough to affect the oxygen content of the lake. It was not until 1953 that it was determined that the problem was caused by a thermal stratification of the lake waters which depleted the bottom of the lake of oxygen by isolating it from the surface. Erie's phosphate and nitrate debit thus outweighed its oxygen credit, creating an unbalanced "budget." "All this means that in using Lake Erie as a dumping ground for municipal and industrial organic wastes and for agricultural fertilizer drainage, there has accumulated—in the bottom mud—a huge and growing oxygen debt."[9]

The picture is complicated by other factors, but Dr. Commoner's conclusion is unequivocal. "In effect, a good deal of the waste poured into Lake Erie has accumulated in the lake bottom as the residue of algae and other living organisms. Thus, instead of Lake Erie forming a waterway for sending wastes to the sea, it has become a trap that is gradually collecting on its bottom much of the waste material dumped into it over the years—a kind of huge underwater cesspool!"[10]

The irony of this situation is that the very wastes being poured into the lake through municipal sewage systems are really needed in a desperate way on the adjacent farmlands. For lack of organic fertilizer and to push production higher (and also because it is cheaper!), the farmers are saturating their soil with inorganic nitrate and phosphate nutrients, which in turn are draining off into the lake to aggravate the aquatic ecology.

Our technological society has thus far successfully polluted more than half of the earth's lakes and rivers to a degree that seriously threatens our way of life and affords little hope of supporting the population growth of the years ahead.

Almost everything we throw away in liquid form eventually finds its way to the sea. Even the airborne pollution sooner or later precipitates into the oceans or the waters that drain into them. It is estimated that we are now adding about a half-million different substances to the earth's waters, the biological effects of which are mostly unknown.

Seeing Consequences

Pesticides like DDT and dieldrin are accumulating in the sea and gradually moving up the food chain to threaten man himself. Highly toxic mercury has already taken its toll in some places and is threatening the fishing industry. In the Mediterranean the level of mercury contamination has reached three times that of the Atlantic, largely due to the industrial waste spewed into the rivers as a by-product of the electrolytic preparation of chlorine and soda.

It was the death of 100 Japanese fishermen in 1953 that alerted the world to the dangers of this maritime pollution. Mercury is eliminated with difficulty from the human body. Normally it takes about a year to totally eliminate, and often only after it has affected the brain and nervous system with insomnia, fatigue, loss of memory and damage to both sight and hearing.[11]

Pollution poses many other threats to marine life, and thus to man, but none is so significant as its capability for poisoning the vegetation which produces much of the world's oxygen. The late Dr. Lloyd Berkner, president of the Graduate Research Center of the Southwest, wrote a paper shortly before his death in 1966 pointing out that 70 percent of the oxygen generated in our biosphere comes from the diatoms, the small free-floating plants which form the basic food of most fish. The other 30 percent comes from land vegetation.

"If our pesticides should be reducing our supply of diatoms or forcing evolution of less productive mutants," he emphasized, "*we might find ourselves running out of oxygen.*" Since we know that the ocean's fish are loaded with pesticides, it is certain that the diatoms are also, for this is how they get into the fish.[12]

Water is the classical symbol of cleanness and life. Jesus used it to represent the supernatural quality of renewal inherent in the gospel. "Everyone who drinks this water will be thirsty again," He said, referring to the well water in His hand, "but whoever drinks the water that I shall give him will never suffer thirst any more. The water that I shall give him will be an inner spring always welling up for eternal life" (John 4:13,14).

If Jesus were revisiting the earth today, I wonder whether He would again choose water as the symbol of His new life. Most water today is polluted and is not always clear and sparkling. Water can also be the purveyor of sickness and death.

Footnotes

1. Raymond L. Nace, *El agua y el hombre: panorama mundial* (Paris: UNESCO, 1970), p. 10.
2. *Ibid*, pp. 28,29.
3. Newspaper clipping, *La Nacion*, San Jose, Costa Rica, October 23, 1974.
4. Newspaper clipping, *The Miami Herald*, Miami, Fla., *circa* October 25, 1974.
5. Claiborne Pell, "The Oceans, Man's Last Great Resource" in *Saturday Review*, October 11, 1969.
6. Edward Goldsmith et al., *Blueprint for Survival* (New York: New American Library, 1974), p. 55.
7. Barry Commoner, *The Closing Circle* (New York: Bantam, 1971-1974), p. 92.
8. *Ibid*, p. 94.
9. *Ibid*, p. 102.
10. *Ibid*, p. 101.
11. Newspaper clipping, *The Miami Herald*, Miami, Fla., October 23, 1974. See also Shigeto Tsuru, "Kogai, Una nueva plaga hace estragos en tres ciudades del Japon" en *El Correo*, UNESCO, July, 1971.
12. Gordon Rattray Taylor, "The Threat to Life in the Sea" in *Saturday Review*, August 1, 1970.

6

OUR DIMINISHING ENVIRONMENT—AIR

My friend Akira Hatori was driving home from his office in Tokyo when his car got caught in a traffic jam while he was negotiating an underpass. Hatori spent three days in the hospital, recovering from carbon monoxide poisoning.

This is not an uncommon occurrence in Japan, where the police patrol cars routinely carry oxygen and gas masks. Japan's relatively small land mass (370,000 square kilometers) supports over 107 million people on the 30 percent of it (112,000 square kilometers) that is habitable. This produces an extremely high population density. In fact, there are historians and sociologists who claim that Japan's elaborate code of etiquette was devised to make it possible for so many people to live in such a small area!

Aggravating Japan's problem is the fact that 72 percent of the people live in cities, and the economy is growing at about 10 percent per year. The nation's automotive vehicles, seven times as many per area as in the United States, cause considerable sickness from the noxious effect of photochemical exhaust fumes. They are directly responsible for 15,000 deaths in accidents per year.

"In Tokyo, Kawasaki and Yokohama," according to *El Demografico*,[1] "some trees lose their leaves four times per year, due, apparently, to the large volume of smoke from automobile exhaust pipes. The birds that have not migrated to less contaminated atmospheres and have not died from insecticide poisoning, suffer from asthma and bronchitis. Accelerated urbanization and the indiscriminate use of insecticides are producing the rapid extinction of some kinds of common birds, as well as some more exotic varieties.

"In Japan," the article continues, "a country known for its respect for natural beauty, growing out of the perfection of its scenery, Nature has tragically lost its equilibrium and is in danger of extinction."

The internal combustion engine and urban life have teamed up to befoul human existence not only in Japan but also in almost every city around the world. During the hours of congestion on our city streets we breathe air containing from 25 to 50 parts per million (ppm) of highly toxic carbon monoxide. The concentration of monoxide permitted in industry is 50 ppm, and the standard of clean air in Florida, for example, is fixed at 10 ppm. Concentrations of more than 1000 ppm are known to produce unconsciousness in an hour and death in four hours.

It is indeed strange that with all the genius and

economic resources available to the petroleum and automobile industries, none of them have come up with pollution-free locomotion. Outside research is not welcomed, and Detroit's record of foot-dragging in the incorporation of anti-pollution techniques in the manufacture of cars is appalling.[2] The only reasonable inference is that the vested interests want to perpetuate the profitable status quo.

Hope for Improvement

Outside of Detroit, a number of hopeful improvements have come to light. One of these, the LaForce automobile engine, was given a test run in November, 1974, by the Delaware Department of Natural Resources and Environmental Control. This modified internal combustion engine was produced by Edward and Robert LaForce. In an American Motors Hornet, driven at thirty miles an hour, it got 31.1 miles to the gallon of gasoline as compared to 18.7 for a Hornet with a standard engine. Besides the 66.3 percent saving in gasoline, the LaForce engine gave off unusually little pollution. Its inventors claim that with additional refinement they can get it up to 100 miles a gallon "real quick."

"I'm extremely hopeful this is going to live up to expectations," said FEA chief John Sawhill in Washington. "It would be a tremendous breakthrough at a time we need a breakthrough."[3]As a matter of fact, further tests have been somewhat disappointing.

While Detroit can certainly be counted on to resist conversion to any other kind of automotive motor, public opinion and economic pressure may be able to produce some important changes. In addition to the Wankel rotary engine, the LaForce engine, and other gasoline-powered improvements on the present automobile power plants, two other possibilities are well worth massive research and development. One is the hybrid combination of an electric motor with a generator powered by a small internal combustion engine. Since the internal combustion engine would run at a constant speed and acceleration would be provided electrically, the pollution would be minimal. The other and possibly the best solution for personal transportation is the steam engine.

"It is our conclusion," noted Dr. Robert U. Ayres in a recent report for the Hudson Institute and the Ford Foundation, "that steam is now the superior alternative on grounds of operating economy, simplicity, intrinsic torque-speed characteristics (which make a transmission superfluous) and the use of lower-octane nonleaded petroleum derivatives such as diesel oil, jet fuel or kerosene. External-combustion engines are also far superior to internal-combustion engines in terms of producing fewer noxious emissions."[4]

Atmospheric pollutants and the problems of smog, smoke, fumes and airborne poisons are almost always associated with cities. Exceptions to this rule would be the radioactive fall-out from nuclear experiments and the pollution caused by aircraft and the ubiquitous automobile. Urban areas suffer simply because they concentrate the greatest needs precisely where there are the fewest resources. Industry and transportation attract more people, compounding wastes and pollution in proportions which nature cannot handle.

The rays of the sun in green areas activate the photosynthetic process in the leaves of trees and plants to produce life-giving oxygen. In the polluted atmosphere of the cities the rays act on the man-made hydrocarbons given off by vehicles and industry to form toxic gases which induce or aggravate human illness.

But photochemical smog, with its nitrogen oxides and hydrocarbons, is only a part of the atmospheric problem in the cities of the industrial world. "In addition," says Barry Commoner, "modern urban air contains: sulfur dioxide and related products of sulfur oxidation; dust originating in furnace ash, industrial operations, and the wearing down of motor tires and asphalt paving by traffic; asbestos particles from automotive brake linings and building materials; mercury vapor from industrial operations; a variety of organic compounds released into the air by combustion and chemical industrial processes.

"A United States Public Health Service study of the air in a number of large United States cities in 1963 recorded, on the average, the presence of thirty-nine different identifiable substances not found in 'natural' air. This list is certainly incomplete," continues Commoner, "for it continues to grow; two major urban air pollutants, mercury and asbestos, have been recognized only in the last few years. The chemical composition of urban dust is still poorly known; only about 40 percent of the mixture of substances that makes up urban dust has thus far been identified."[5]

Air Pollution and Health

Obviously, it is not easy to identify any airborne pollutant as the cause of a particular sickness. There are too many other factors involved. A particular form of lung cancer, however, has been traced to asbestos fibers. And certain metals, such as cadmium, have been shown to contribute to heart disease. Beyond these cases, not much can be proved "scientifically." "However," Commoner points out, "statistical studies do show that people living in urban polluted air experience more disease than people living in less polluted, usually rural, air . . . . The conclusion that *can* be firmly reached is scientifically crude, but meaningful: polluted air makes people sicker than they would otherwise be and hastens their death."[6]

In global terms, the concentration of carbon dioxide is especially serious. The CO_2 content of the atmosphere has been increasing each year at the rate of 0.2 percent since 1958. The report of the Study of Critical Environmental Problems (SCEP) at MIT points out that the trend towards depleting the remaining stands of original forests will further reduce the capacity of the ecosphere to absorb CO_2. SCEP predicts that by the year 2000 the earth will have experienced an 18 percent increase of CO_2. This might raise the mean surface temperature of the globe by two degrees centigrade.[7]

No one is quite sure exactly how much oxygen depletion and carbon dioxide pollution the earth's atmosphere can take. But by the time we find out it may be too late. And the quality of our human existence will have been irreversibly diminished.

Footnotes

1. *El Demografico,* Bogota, Colombia (No. 7) Diciembre, 1973.
2. See Barry Commoner, *The Closing Circle* (New York: Bantam, 1971-1974), pp. 66-68.
3. Newspaper clipping, *The Miami Herald,* Miami, Fla., November 25, 1974.
4. David Rorvik, "The Transport Revolution" in *Project Survival* by Geoffrey Norman (Chicago: Playboy Press, 1971), pp. 251,252.
5. Barry Commoner, op. cit., pp. 72,73.
6. *Ibid.,* p. 75.
7. Edward Goldsmith et al., *Blueprint for Survival* (New York: New American Library, 1974), pp. 63,64.

7

OUR DIMINISHING ENVIRONMENT—ENERGY/METALS

"The bulk of our energy requirements today is met by fossil fuels, which like metals are in short supply," states *Blueprint.*

"At present rates of consumption, known reserves of natural gas will be exhausted within 35 years, and of petroleum within 70 years. If these rates continue to grow exponentially, as they have done since 1960, then natural gas will be exhausted within 14 years, and petroleum within 20.

"Coal is likely to last much longer (about 300 years)," *Blueprint* continues, "but the fossil fuels in general are required for so many purposes other than fuel—pesticides, fertilizers, plastics, and so on—that it would be foolish to come to depend on it for energy."[1]

The world was shocked when the Arab countries

suddenly upped the price on crude oil. It should not have been surprising. They simply became aware of the facts we have cited, and realized that at current prices for the forty or so years of continued production anticipated for Arab oil wells, they would not be able even to finance their own industrial and economic development. So they raised the price.

As the Club of Rome pointed out,[2] this was probably a blessing in disguise. It has brought back into competition petroleum sources that were previously too costly to exploit. It has accelerated exploration for new oil. And it is spurring research for other energy sources.

From the standpoint of the environmentalist, the disturbance caused by the Arab maneuver has its silver lining. The use of fossil fuels exhausts the earth's known reserves, and is also ecologically a most disruptive source of energy. It is almost as threatening as the use of nuclear power.

The amount of energy used in the developed, industrial nations borders on the shocking. In the United States every year each person consumes the amount of energy produced by the combustion of about 12 tons of coal. This compares with about 10 tons in Canada and 2.75 tons in Venezuela, with 1.35 in Mexico and .65 in Colombia. And in Ethiopia, any peasant could carry his year's supply of energy in a sack over his shoulder. It amounts to the equivalent of 36 kilograms, or 80 pounds, of coal.

Obviously the use of energy is accelerated by affluence. During a 20-year period from 1946-1966 in the United States, the population increased by 43 percent. The use of energy jumped by 100 percent. It is estimated that by the end of the century the U.S.

population will have increased by about 30 percent. The use of energy will have increased by 400 percent to 500 percent.

The explanation of these statistics lies in the technological development of the nation as well as in its growing affluence. Not only does the average American travel more and heat a larger home, he makes use of technological substitutes for natural products. The synthetic fibers of his wearing apparel, the plastics used in packaging his consumer goods, his processed foods and the mechanized, chemical techniques used to grow them consume energy of one kind or another.

Nuclear Power

So the heavy use of fossil fuels is contaminating and is exhausting known resources. What about nuclear power?

Environmentalists like Barry Commoner are very skeptical about nuclear power plants. The emission standards first issued by the federal government were criticized by many scientists as being too lenient and were adjusted somewhat in 1971 after years of strong resistance on the part of the Atomic Energy Commission. Even now, the State of Minnesota is involved in controversy with the AEC because it has set standards higher than those of the federal controlling agency.

Despite "accidents" such as the deaths of crew members of the Japanese fishing vessel, "Lucky Dragon," and the appearance of strontium 90 in the milk of breast-fed babies in the Middle West, there is a common conviction that the new atomic knowledge is sound, the technology competent, and the power irresistible.

"The first twenty-three years of the atomic age tell us that this belief is deeply, tragically, wrong," Commoner insists. "Isolated on a Pacific island or confined to the grounds of a power plant, nuclear energy is a success. It works: it vaporizes the island; it sends electricity surging out of the power plant." Neither the island, power plant, nor anything else can be separated from the "thin, dynamic fabric" of the environment that envelops the planet. "And once power from the split atom impinges on the environment, as it must, we discover that our knowledge is incomplete, that the new technology is therefore incompetent and that the new power is thereby something that *must* be governed if we are to survive.

"This, it seems to me," concludes Commoner, "is the meaning of the first environmental encounter of the new age of technology. Our experience with nuclear power tells us that modern technology has achieved a scale and intensity that begin to match that of the global system in which we live. It reminds us that we cannot wield this power without deeply intruding on the delicate fabric that supports us. It warns us that our capability to intrude on the environment far outstrips our knowledge of the consequences. It tells us that every environmental incursion, whatever its benefits, has a cost—which, from the still silent testimony of the world's nuclear weapons, may be survival."[3]

Given the high cost of energy derived from the fossil fuels, it does not seem likely that even the most dedicated of ecologists will be able to stay the advance of the nuclear technology. The only spontaneously fissionable source of nuclear energy is uranium 235. This is likely to be in extremely short supply by the end of the century.[4] The future of nuclear power rests, therefore, with the development of complete breeding systems which utilize the neutrons from the fission of uranium 235. These convert non-fissionable uranium 238 and thorium 232 into fissionable plutonium 239 and uranium 233, respectively. Should the development of this technique prove to be successful, economically viable, and ecologically harmless, our energy needs will probably be taken care of for the next 1000 years or so. By then other technologies (e.g., deuterium-deuterium) may have evolved.[5]

Apart from radioactive wastes and combustion pollution, heat pollution is one of the most serious energy problems. At the present time in the United States, electricity provides 10 percent of the power actually used by the consumer. Its generation accounts for 26 percent of the nation's gross energy consumption. By the year 2000 electricity will be providing 25 percent of the consumer power, but will itself be consuming between 43 and 53 percent of the gross energy consumption. At that point, half the energy produced will be in the form of useful work and half in the form of waste heat from power stations.[6]

If we knew how much waste heat our ecosphere can absorb it would help to determine the direction and focus of our industrial expansion. Unfortunately, our knowledge is limited. We do know that if the projections cited by *Blueprint* are correct, within 99 years man will be producing waste heat equal to 17 watts per square foot in the United States, compared to the average 18 or 19 watts received from the sun. Concludes *Blueprint*: "Clearly, well before this point, energy consumption will be limited by the heat tolerance of the ecosphere."[7]

Alternate Sources

These and other considerations give added importance and impetus to contemporary research in other areas of energy production. Within a short time solar water-heating systems and solar furnaces for homes will undoubtedly become available. Solar air-conditioning units are being tried on an experimental basis in southern Florida. The prospects are not yet promising. Solar energy is, of course, the source most harmonious with nature's ecosystems.

Hydroelectric power plants in countries fortunate enough to have harnessable rivers continue to be one of the best sources of energy, Sometimes these projects backfire as in the spread of a serious disease, schistosomiasis, in the irrigation ditches of the Aswan Dam in Egypt. Sometimes they fail to measure up to expectations because of irresponsible forest management. This happened with the hydroelectric power project dam in Colombia which filled up with mud from the eroded mountains only twelve years after its completion. In western Pakistan the 600-million-dollar Mangla Dam, originally calculated to serve for 100 years is silting up in half that time and for the same reason. In every case, the damming of rivers causes ecological disturbance, although nature can sometimes adjust to the changes. Even in hydroelectric projects, environmental factors must not be overlooked.

A number of ingenious ideas for power plants are being seriously explored. One suggests harnessing the relatively constant eruptions of volcanos, like

Arenal and Vesuvius. Another proposes to force the tidal pressures at Mont-Saint-Michel to turn the turbines of an electric generator. Still another suggests using the ocean's thermal currents in a floating power station to alternately heat and cool a gas like freon. Many of these and other proposed alternatives pose less of an ecological hazard than the present combustion of hydrocarbons and the use of nuclear power.

In any consideration of non-renewable resources, metals must be carefully studied. Out of the sixteen major metals, at present rates of consumption, all known reserves of ten of them will be exhausted within 100 years. If rates of consumption continue to increase exponentially at the rate they have done since 1960, these reserves will be totally depleted within fifty years, as will those of all but two of the remaining six metals.[8]

Admittedly, new sources of some metals may be found in the interim. And recycling may be able to stretch our reserves somewhat. The picture remains ominous. Particularly critical are the low reserves of silver, gold, copper, mercury, lead, platinum, tin and zinc.[9]

A simple solution would seem to be the use of synthetic substitutes for scarce metals. Most of the synthetics are themselves derived from scarce materials, however. Petroleum, for example, from which many valuable synthetic polymers are derived, will probably run out during the lifetime of those born this year. After the year 2000 it will be increasingly scarce and expensive.

As the costs of producing energy go up, and the richness of ore deposits go down with the exhaustion of the richer lodes, metals will soon disappear or price themselves out of the industrial market. "Only those acrobats of the imagination who argue that, come what may, technology will find a way, believe that problems such as these can be solved in any way save by diminution of consumption."[10]

Expensive Waste

The squandering of valuable resources in trash in the developed countries, particularly, is scandalous. Less than one percent of the trash from the United States' cities and towns is recycled. According to the U.S. Bureau of Mines, the half-million tons of fly ash produced from burning America's refuse each year could yield 150,000 pounds (67,000 kilograms) of reclaimed silver.

Besides retrieving valuable metals, U.S. government experts also claim that shredded, burnable trash could produce one quadrillion BTU's of energy a year, about one-third of what is expected from the oil flowing through the Alaskan pipeline.

Every American produces 5.5 pounds (2.46 kilos) of trash per day, and the annual United States trash "production" is enough to fill the Panama Canal four times. Most of this ends up in dumps and incinerators, on roadsides, vacant lots, parks and beaches. Waste-making cannot be allowed to continue for obvious reasons. Soon we will have turned most of our precious metal reserves into unredeemed waste.

This situation must surely say something about the life-style of our affluent cultures. In 1973 worldwide automobile production totalled 30,525,006 vehicles. Almost half of these cars weighed more than 1500 kilos (3,360 pounds). If each of these bigger cars had weighed just 200 kilograms (500 pounds) less, the savings in raw materials alone could have totalled an estimated 3,000,000 metric tons.

If by reduced weight and other engineering improvements, each of the cars could have attained a mere 15 percent increase in fuel economy, savings in gasoline would have totalled 4 billion liters, or approximately 888 million gallons.

A recent advertisement by Fiat describes the situation succinctly: "We're not only running out of energy, we're running out of earth."

Footnotes

1. Edward Goldsmith et al., *Blueprint for Survival* (New York: New American Library, 1974), p. 115.
2. Club of Rome, *Mankind at the Turning Point*, quoted by *Time* magazine, October 21, 1974.
3. Barry Commoner, *The Closing Circle* (New York: Bantam, 1971-1974), p. 61.
4. Edward Goldsmith et al., op. cit. p. 115.
5. *Ibid*, p. 116.
6. *Ibid*, pp. 116,117.
7. *Ibid*, p. 114.
8. *Ibid*, p. 13.
9. *Ibid*, p. 115.
10. See "Action Line," *The Miami Herald*, Miami, Fla., November, 1974.

8

OUR DIMINISHING ENVIRONMENT—FOOD

This is where the crunch comes—food!

Here all the lines of population growth and resource depletion converge. Man can exist without plutonium or the internal combustion engine or electricity. But food is essential.

Inadequate quantities and varieties of food produce malnutrition exposing men and women and children to disease and premature death. Nearly half of the population of the Lesser Developed Countries (LDC's) suffers from undernourishment.[1] They are prey to every sort of sickness, particularly to kwashiorkor, rickets, beriberi and pellagra.

After an individual has lost about a third of his normal body weight he is technically starving. He begins to burn up his own body fats, muscles and tissues for fuel. Once his weight loss exceeds 40 percent of normal body weight, death is almost inevitable.

Most adults can come close to starvation and survive, but children remain scarred for life. When their legs have been bowed by rickets there is no way to straighten them out. At least 80 percent of brain growth occurs between the time of conception and the age of two, which means that babies who are starving or whose mothers endured starvation during their pregnancy are subjected to irreversible privations. Brain growth which has not occurred during the first two years will never take place at all. Hunger is condemning thousands of infants around the world to the incurable shadows of mental retardation.

Starvation of these tragic proportions is at this moment prevalent in the ten nations of the African Sahel, as well as in India, Bangladesh and northeastern Brazil. Another dozen nations are in a precarious situation with national averages of per capita calorie consumption lower than estimated daily requirements. These latter include several Latin American countries: Haiti, Bolivia, El Salvador and Ecuador, which are classified as "potential problem countries."[2]

To these alarming situations should be added the large segments of the population of practically every other nation which are chronically undernourished but whose calorie deficits do not bring the national average below the danger level. Colombia, a nation of 23.7 million inhabitants, boasts national average calorie consumption per capita each day of 2,200. This is a bare minimal average for a tropical country. At the same time, 77 percent of the population is undernourished and 30,000 deaths per year of children under six years of age are attributed directly or indirectly to malnutrition.[3] The national average of calorie intake is thus a deceptive statistic. Like the GNP (gross national product), it is sustained by the excessive calorie consumption of the affluent minority.

"The stark truth is that man's ability to produce food is not keeping pace with his need," summarizes the *Declaration on Food and Population.*[4] "Despite efforts by governments and the international community to solve world food problems, more people are hungry today than ever before."

The reasons for this are not hard to identify.

The crop deficiencies of 1972 and 1973 have highlighted the world's precarious food situation. Droughts in the African Sahel and in the Indian subcontinent, plus similar although less severe situations in the North American Middle West and elsewhere, produced harvests below those of previous years and substantially below the demands of a growing and more affluent population.

Coincident with the drought was the temporary disappearance of the Peruvian anchovieta, an important source of animal feed and fertilizer, and a general world-wide decline in the catch of fisheries.

On the heels of these misfortunes, the legendarily inexhaustible grain reserves of the United States were suddenly depleted by a massive sale of wheat to Russia. Russia's crops had also been affected by the drought. Her affluence caused her to import grains rather than to slaughter cattle, as she had done during the last grain crisis.

Then came the knock-out punch.

A sudden and unanticipated escalation in world prices of petroleum hit the entire food industry in the midriff. Fertilizer, already in short supply, became a luxury. The price of pesticides, also petroleum derivatives, shot up. Mechanized agriculture became much more costly, as did transportation and every other operation essential to the production and distribution of food.

As a result, the world's reserves of grain have reached a 22-year low, equal to about 26 days' supply, compared with a 95-day supply in 1961. So precarious is the food situation that almost anything —bad weather, a longshoremen's strike, an oil embargo—could precipitate apocalyptic famine in some part or parts of the globe.

Food Losses

Other perennial problems further complicate the situation. At least 25 percent of the world's food disappears between the field and the table. Especially in the LDC's, food is often inadequately warehoused and is subject to damage by rats, insects, fungus and mildew. Another much more complicated problem is the whole matter of land ownership and incentives to put arable land into production. In some countries the reform of land ownership is more important than its fertilization.

Many experts have looked to technology and the "green revolution" to solve the world's food needs. With increasing success in the last decades, agrobiologists have bred new varieties of grains that produce vastly larger crops on shorter and stronger stalks to withstand the wind as well as the weight of the grain at harvest time. Improvement has been registered in all the common grains with the most dramatic in rice and wheat strains. The agricultural comeback of India and other Asian countries after the 1966-67 droughts is attributed largely to the "miracle" grain seeds.

Unfortunately, in the genetic evolution of these new strains, the plants tend to lose their resistance to pests and diseases along the way. Also, they require considerably more water than the older grains and are dependent upon increasing amounts of nitrogen fertilizers to produce maximum crops. Without massive fertilization programs and large-scale use of pesticides, the harvest is not likely to exceed that of the older, traditional seeds.

The current prospects for a bumper crop in India are not good because the nation can afford in 1974 to import only half the fertilizer needed for a maximum yield. As a result, India anticipates an 8- to 10-million ton drop in production of wheat and rice. Basic to this situation is the fact that the world price of nitrogen fertilizer jumped from $.11 per pound in 1972 to its current 1974 level of $.25, an increase in two years of 127 percent.

Apart from the scarcity and high price of petroleum, the world's capacity to produce fertilizer in the areas where it is most needed is presently far from adequate. Each new factory will cost up to $100 million to build and will consume enormous amounts of energy. Dr. Raymond Ewell, professor of Chemical Engineering at State University at Buffalo, New York, estimates that any country needs one additional comprehensive fertilizer plant for every additional 6 million people. This would require India to build two and one-half new plants per year, and they simply aren't building them that fast.

On a global basis, "the total world investment required right now per year $8 billion to keep up with demand. And this figure will grow to about $12 billion per year by 1980 . . . . The present rate of investment to the fertilizer industry is more on the order of $4 or $5 billion, so we're going to fall progressively behind."[5]

For these economic reasons, the green revolution is a subject of controversy. Many eco-biologists and organic agriculturalists discredit it on the basis of damage to the environment. "Agribusiness is founded on several technological developments," states Barry Commoner, "chiefly farm machinery, genetically controlled plant varieties, feedlots, inorganic fertilizers (especially nitrogen), and synthetic pesticides. But much of the new technology has been an ecological disaster; agribusiness is a main contributor to the environmental crisis."[6]

Potentially the most dangerous of the agribusiness technologies is the massive application of inorganic, mostly nitrogen, fertilizers. Once started, the soil gets "hooked" on nitrogen and requires constantly increasing quantities to produce the oversize crops. Probably this is because the inorganic chemicals somehow cause the soil to lose its capacity for absorbing organic nitrogen. This massive use of fertilizers causes enormous wastage and drain-off which pollutes the surface water systems. Pollution endangers the health of infants from methemoglobinemia and destroys the algae and oxygen balances in streams and lakes.[7]

Affluence as a Cause

Basically, the world-wide shortage of food can be attributed to overpopulation. But the problem is not that simple. Distribution of existing resources is radically out of kilter. Affluence is one of the major contributors to the existence of famine in many parts of the world. As living standards in both developed and developing countries rise, the affluent citizens take a constantly bigger bite out of the world's slice of bread. Because it automatically deprives starving children of that much more food, agronomist Rene Dumont calls this "cannibalism."[8]

Residents of Tucson, Arizona, were found recently to be wasting between $9 and $11 million worth of edible food annually. This constitutes 9 percent of the food by volume which comes into Tucson homes, excluding liquids. And if the waste food ground up in garbage disposal units were added to the above, if would come closer to an estimated 15 percent of total food consumption.[9] Food waste from uneaten portions in restaurants is calculated to be even higher. Add to this the millions of tons of food fed to pets, and the drain of affluence on the world's food stores becomes staggering.

Even more serious is the habit of affluent people to consume increasing amounts of feedlot-fattened beef. In most underdeveloped countries the average citizen eats 400 pounds of grain per year, 360 pounds of it in the form of grain, bread or gruel, and the balance in animals and fowls partially nourished on grain. In the United States and wherever beef is fattened in feedlots, the average citizen consumes five times as much—a ton of grain, most of it in the form of grain-fed beef, pork and chicken. Twenty pounds of grain in the feedlot will produce one pound of beef on the hoof. In times of famine, this is worse than an anomaly.

The problem of food is a complex one, involving

in its solution factors related to economics, demography, sociology, ecology and human attitudes, biochemistry and agronomy.

In his first appearance before the United Nations, U.S. Secretary of State Henry Kissinger proposed a World Food Conference. The political and social leaders of the nations could study the matter in its total perspective and come up with solutions, both long-range and immediate, to the fast-spreading menace of hunger, malnutrition and famine. Accordingly, in November, 1974, about 1000 delegates from some 100 nations and a dozen international organizations gathered in Rome to grapple with the universal issues of the food problem.

The results hardly measured up to the fanfare. But the conference did register some positive accomplishments. Most noteworthy:

1. The delegates agreed to set up a World Food Council under the United Nations. It will be based in Rome, and will channel to the needy countries of the world food aid as needed and available. At the same time it will funnel investment funds for the development of agriculture.

2. It was also agreed to set up an international grain reserve, with grain stocks to be held by individual nations.

3. Food donor countries promised to aim at giving away 10 million tons of grain per year over the next few years.

4. A consensus was reached on an "early warning system" to spread information about the state of world harvests and hunger.

Basic and immediate remedies were in short supply, however. President Ford rejected a last-minute plea for an immediate doubling of food aid to starvation areas overseas. This would have escalated domestic food prices beyond the reach of American housewives, a justification that could hardly be expected to satisfy Third World critics of U.S. policies, despite the fact that the United States has furnished 46 percent of all food relief received by developing countries over recent years.

The fact is that immediate solutions to the food problem are almost too costly to contemplate. *Time* magazine claims that "it is scientifically and technologically possible to feed a world population several times its present size. Yet this projection ignores two serious limitations: the huge cost and the problem of convincing citizens of wealthy nations that they must sacrifice to help those in poor countires."[10]

To irrigate 57 million extra acres of farm land—a 25 percent increase over present irrigated acreage—would cost 3.5 billion annually for the next eleven years. It would take another 5 billion per year to set up an agricultural development fund which the FAO has been requesting. We have already seen how much it would cost to expand fertilizer production to meet the estimated demands. The price of bringing new land under cultivation is staggering. An approximate 10 percent increase in the world's arable land—adding 400 million acres—says *Time*, would cost at least $400 billion and might run to $1 trillion or more!

A careful look at the facts leaves this observer glum. The cost and complexity of a truly global solution to our planet's food problem is staggering beyond comprehension. And yet, we must agree with the *Declaration on Food and Population* that "food . . . is the most critical of the pressures on the world today. It is the greatest manifestation of world poverty, which has many aspects. The absolute number of desperately poor is far greater today than ever before in history .... We repeat, food is crucial because literally tens of millions of lives are suspended in the delicate balance between world population and world food supplies ....

"In the name of humanity," the Declaration concludes, "we call upon all governments and peoples everywhere, rich and poor, regardless of political and social systems, to act—to act together—and to act in time."

Footnotes

1. Robert McNamara, *El Demografico*, Bogota, Colombia, October, 1973.
2. *Time* magazine, November 11, 1974.
3. *El Demografico*, January, 1974.
4. The *Declaration on Food and Population* was a "call to governments and people for action" signed by more than 2,500 "distinguished citizens from over 100 countries." It was issued to underline the significance of the two world conferences convened under United Nations' sponsorship in 1974: that on "population" at Bucharest in August, and that on "food" at Rome in November.
5. Interview with Raymond Ewell and Norman Borlaug, "The Shrinking Margin" in *Ceres* (38), FAO Review and Development, March/April, 1974, p. 56.
6. Barry Commoner, *The Closing Circle* (New York: Bantam, 1971-1974), p. 146.
7. *Ibid*, pp. 79,102.
8. Rene Dumont, "Population and Cannibals" in *Development Forum*, Centre for Economic and Social Information / OPI, Geneva, September/October, 1974.
9. Newspaper clipping, *The Miami Herald*, Miami, Fla., November 24, 1974.
10. *Time* magazine, November 11, 1974.

9

WHAT ARE WE DOING WRONG?

I went through the security check at the Miami airport and got around to the other side of the hand-baggage machine in time to see on the television screen a full-size X-ray negative of my attache case, complete with paper clips, spiral bindings and pocket computer. Everything was there, clearly depicted, without confusion.

In much the same way, as our civilization moves through the current crisis in environment, even an untrained observer can now detect forms and flaws in its structure that were previously only imagined.

What are we doing wrong? From an ecological standpoint, just about everything! But a few factors stand out sharply.

Most conspicuous is the technological flaw.

The year 1946, the end of World War II, was a watershed year in science. Mankind has "advanced" farther in technology since 1946 than in all previous history. Just try to think of the everyday technological paraphernalia of life which were unknown or were restricted to research laboratories before World War II. Jet propulsion, the transistor and solid-state electronics, color TV, computers, plastics, synthetic fibers, rocket propulsion, space travel, miracle drugs and open-heart surgery are but a few. We are living in the greatest technological explosion mankind has ever experienced.

Coincidentally, the period since 1946 has been the period of greatest environmental pollution. Barry Commoner points out that the population of the United States grew 42 percent during the twenty-five years following World War II, and the gross national product increased at a proportional rate (126 percent). At the same time, a very pronounced change took place in the consumption habits of the American public. The production of nonreturnable soda bottles jumped about 53,000 percent during that time, and synthetic fibers some 5,980 percent. Other fast-growing products included mercury (for chlorine production and for use in mildew-resistant paint), air-conditioner compressor units, plastics, nitrogen fertilizers and electric appliances.[1]

Production and consumption of food, textiles and clothes, household utilities and basic metals all increased in approximately the same rhythm as the population. In other words, per capita production and consumption of these items remained more or less constant. The consumer explosion had taken place in the products of the new technology.

An examination of each phase of technological growth would show that the new technologies have had a greatly increased impact on the environment. "This pattern of economic growth," says Commoner, "is the major reason for the environmental crisis. A good deal of the mystery and confusion about the sudden emergence of the environmental crisis can be removed by pinpointing, pollutant by pollutant, how the postwar technological transformation of the United States economy has produced not only the much-heralded 126 percent rise in GNP, but also, at a rate about ten times faster than the growth of GNP, the rising levels of environmental pollution."[2]

Now we are at the end of our ecological rope. Although the technocrats have been responsible for the greater part of the mess we are in, we still tend to look to technology for solutions to our global problem. Is it logical to expect that technology will reverse its mental mind-set and present us now with the solution?

Gene Marine, in an interesting article on "The Engineering Mentality,"[3] accentuates the futility of such an expectation. "Point out to a technician that a river sometimes floods its lowlands (or that there's a market for hydroelectric power in a nearby town) and he builds a dam. Point out to him that the dam will eliminate the salmon run on the river and he builds a fish ladder and artificial gravel spawning pits. Point out to him that the lake behind the dam will drown a small village on an Indian reservation and, at best, he'll call another technician to build a model city for Indians.

"What he will *never* do," affirms Marine, "is reconsider the idea that he ought to build the dam in the first place. He won't do that because he can't; the engineering mentality just doesn't work that way." As a technologist his job is simply to solve the problem presented to him.

Technologists are essential to our modern world. But when political decisions become problems needing a technological solution, many human and social values are lost.

"Give the technicians a difficult foreign-policy question as in the Dominican Republic,"[4] suggests Marine, "and their answer is to call it a problem and to solve it by invading the country and interfering in its political processes. If this creates new problems throughout Latin America for twenty years afterward—well, we'll deal with those one at a time as they come up."

The technological flaw is the result of specialized scientists, who suffer from tunnel vision. They zero in on one problem at a time without taking adequately into account the environmental impact of their proposed solutions.

Look at our urban industrial sprawl. It is a technological "solution" to social and economic problems. To provide man with more comfort and leisure a new tool is invented and a factory set up to manufacture it. The new industry needs manpower, so dwellings come into existence. Supporting and supplementary industries are needed. The city grows. Transportation of raw materials and of manufactured goods requires railroads and shipping yards. Transport volume grows. Highways are built. The cost of land in the city center goes up, and the only solution is high-rise office and apartment buildings.

Meanwhile, every problem of human ecology is

increasing dramatically. Power consumption is massive. New plants must be built. Sewage and garbage control require constantly larger facilities as do fire-fighting, policing, schools, hospitals, street-sweeping services, etc.

People Become Statistics

We have a modern tower of Babel in which persons become statistics, people become units, and society is dehumanized. There remains no neighborliness, no mutual concern. Our original objective—to better man's living conditions—is forgotten. The inhabitants of the large city live in isolation from each other, from nature, and from God.

Two small indicators of the situation in our dehumanized metropolis: one, it costs far more to police a city of one million inhabitants than it does to police ten cities of 100,000; two, a much smaller percentage of the large city dwellers are found in church on Sunday morning than in smaller towns and rural areas. These facts say something about the quality of human existence in the industrial urban concentrations.

The urban sprawl is a good example of a technological end product. So is the household detergent.

Ordinary soap is made from natural products—such as palm oil and lye—with very little consumption of energy or polluting by-products. After it is used, the soap washed down the drain has little or no effect upon the aquatic ecosystem.

By contrast, detergents are synthesized from petroleum products with a much greater expenditure of energy. And to the cleansing agent are added a number of substances such as chlorine, whose production requires the use of mercury. The impact on the environment from the manufacture of detergents is many times that of soap.

The most important contrast comes after the detergent is used and washed down the drain. Whereas soap has been used everywhere in the world for thousands of years without any serious ecological problem, detergents have chalked up a very bad environmental record in the twenty-five years of their common usage. At first they were not biodegradable, and foam began to appear on streams and to come out of down-river faucets. When this was corrected, other chemical reactions such as the production in aquatic systems of carbolic acid were detected. The phosphate water-softening additive has stimulated overgrowth of algae in countless surface waters.

Detergents continue to reign in the market place, however, because technology has made it possible to produce them more cheaply than soap. The profits of the manufacturing company outbalance the loss to the environment, in the minds of the technologists and those who pay their salaries.

We come reluctantly to the conclusion that technology is perhaps competent to give an assist at each step along the journey of life, but that we should not look to the technologists for direction in determining our course. We need to depend more on the generalists than on the specialists.

There are some technological breakthroughs which would do much to help us solve the overwhelming problems of food and environment we face today. They might buy us time for more intelligent planning for whatever future remains to us. If the technologists could make every cow have twins instead of only one calf per year, this would vastly augment our sources of protein. If they could create some kind of fodder from sewage solids, or manufacture proteins by the cultivation of water fleas in sewage disposal plants, or breed a variety of soybean which would yield double the present crop, we could buy time to face up to our present food crisis.

However, the Club of Rome's *The Limits to Growth* is careful to point out that the best technology can do for us is to slightly extend our finite limits. "Applying technology to the natural pressures that the environment exerts against any growth process has been so successful in the past that a whole culture has evolved around the principle of fighting against the limits rather than learning to live with them."[5] Until recently this concept has been reinforced by the apparent immensity of the earth and its resources.

But now things are changing. The exponential growth curves are adding millions of people and billions of tons of pollutants to the ecosystem each year. The conclusion: "Our attempts to use even the most optimistic estimates of the benefits of technology in the (computer) model did not prevent the ultimate decline of population and industry, and in fact did not in any case postpone the collapse beyond the year 2100."[6]

Distribution of Wealth

Another conspicuous flaw is in our economic structure.

The famine encircling our globe is evidence that a capitalist structure of free enterprise produces great wealth for many people who are sufficiently endowed with intelligence, opportunity and motivation. It does not provide adequately for the underdeveloped, the underprivileged and the undermotivated inhabitants of the earth. This is because capitalism and the amassing of personal wealth must be balanced with altruism (genuine philanthropy, public-spiritedness, unselfish concern for others) if it is not to become unbridled tyranny and imperialism.

There is still a great deal of positive social dynamic in capitalism. In the United States the tax laws, the philanthropic foundations and the underlying Christian work-ethic all contribute to a social structure that has served us well for many years. Recently, the hungry minorities in Appalachia, the black ghettos and the shantytowns of the deep South have become festering indications that all was not well. As the specter of world-wide famine haunts subequatorial Africa, southeast Asia and Latin America, we can see clearly the outline of our failure.

Developmentalism can never be enough. It must go hand-in-hand with social justice. Most foreign aid has simply made the rich richer and the poor poorer. Unless development can, as Barbara Ward says, be "thrust down to the grass roots," it will be wasted energy.

Even the "green revolution" in India has backfired in this regard. The miracle seeds were designed to require more farm labor. But the wealthy farmer has been able to mechanize his farm, reduce labor, speed the harvest-to-planting turn-over, and rake in the profits. One landowner gloated that he had netted $100,000 in one year by commercializing the green revolution. Many Indian farmers could not afford to pay for the fertilizer the new seeds required. The man who needs help can't get it, and the one who doesn't need it, grows wealthier.

There is even some basic question as to whether the profit motive can create or permit an adequate concern for the preservation of the environment. The record of private industry has not been good. It has been our biggest polluter. The public is paying the bill in environmental destruction and depletion. Industry would change price structures and attitudes drastically if it were required to figure in the cost of correcting the damage done to the ecosystem by industrial waste and effluents. This must be done. It is ridiculous to think that the common heritage of all men should be demolished or diminished for the personal profit of a few industrialists.

Perhaps our whole capitalistic concept of personal possession of property should be re-examined. To whom do the resources of nature belong? Whose are they to enjoy? Who is responsible for their replenishment? Karl Marx said: "Private property has made us so stupid and partial that an object is only ours when we have it, when it exists for us as capital or when it is directly eaten, drunk, worn, inhabited, etc., in short, utilized in some way."[7]

In criticizing the underlying concepts of capitalism, developmentalism and consumerism, I don't mean to imply that Communism or Socialism is necessarily the answer. I simply want to make clear that an uncorrected capitalism is exploitive and tends to be ignorant or arrogant towards environmental responsibilities.

"I would say that the pollution-environment-resource problems of socialist and capitalist societies are essentially identical," observed Dr. Paul Ehrlich.

"The Soviets and the Chinese are just as bad or worse in regard to their environment and resources as we are. In fact, Marxism is conceptually worse, because Marx, being an enemy of Malthus, found it unthinkable that an infinite number of people couldn't be supported if the Communist system were running the world. So, as bad as our government is, it would be worse if it were Marxist. It's not a matter of socialism or communism versus capitalism; it's a matter of the exploitive economy having to become a conserving, recycling economy."[8]

One thing is sure. There are going to have to be a lot of changes. Changes in the attitude of industries, changes in the posture and effectiveness of government officials, changes in the priorities of towns and communities, changes in the life-styles of individual citizens may avert a doomsday crash. And yet, there is nothing within the system which would give hope that enough of these changes can and will take place.

Our situation is analogous to that of an automobile speeding along the express lane of a superhighway. There's no way of getting off until you get to the end of the road. Maybe we can slow down a bit and ease over into a slower lane. But the highway is one of limited access and exit.

There's just no way we can make a U-turn.

Footnotes

1. Barry Commoner, *The Closing Circle* (New York: Bantam, 1971-1974), p. 141.
2. *Ibid*, p. 144.
3. Gene Marine, "The Engineering Mentality," excerpt from *Project Survival* by Geoffrey Norman (Chicago: Playboy Press, 1971), pp. 211,212.
4. *Ibid*, p. 217.
5. Dennis L. Meadows et al., *The Limits to Growth* (New York: Universe Books, 1972), p. 156.
6. *Ibid*, p. 152.
7. Quoted in Barry Weisberg, *Beyond Repair, the Ecology of Capitalism* (Boston: Beacon, 1971), p. 73.
8. Paul Ehrlich interview, *Project Survival* (Chicago: Playboy Press, 1971), p. 108

10

THINGS TO BELIEVE IN

It has become fashionable for ecologists to find the sources of our environmental crisis in the Judeo-Christian mind-set. Superficially, it is an easy case to make.

Professor Lynn White, Jr., of U.C.L.A.[1] sees science and technology as being basic projections of Christianity's "natural theology." He says Western Christianity is the most anthropocentric religion the world has known. In it, man is created in the image of God and reflects God's transcendence in his dominion over nature. This creates a dualism which has provided a rational theological context for the exploitation of natural resources by man for his own ends.

The victory of Christianity over paganism was the greatest psychic revolution in the history of our culture, White maintains. By destroying pagan animism,

Christianity made possible the exploitation of nature with total indifference to the intrinsic value of the natural objects themselves.

Our science and our present technology are so colored by a "Christian arrogance" towards nature, he concludes, we cannot look to them for solutions to our ecological problems. Rather, since the roots of the crisis are religious, the remedy must also be essentially religious.

The solution which White proposes is a revival of the philosophy of St. Francis of Assisi, who respected and loved every living thing because it had come from the hand of God the Creator. St. Francis' kinship with the world of nature provides a religious corrective to the "Christian arrogance" to which so many of our environmental ills can be attributed.

Most ecologists today seem to favor pantheism. As they discover the interdependence of life's ecosystems they move very easily into the position that Reality is the totality of these systems and is found in each part as well as in the whole.

This, in fact, is the posture assumed by sociologist Richard L. Means in an article headed "Why Worry About Nature?"[2] He takes his cue from the hippies who have swarmed into Zen Buddhism. He does not advocate a romantic, individual sort of identification with nature, but finds the "technological arrogance" springing from Calvinism and deism to be basically immoral. And to him the only reply to the ecological crisis seems to be what we would call pantheism.

Dr. Schaeffer in his *Pollution and the Death of Man*[3] comes to the conclusion that the natural theology of the Christian faith has indeed given rise to science and technology and that these stepchildren have been all too arrogant and self-seeking in their exploitation of nature.

But this, asserts Dr. Schaeffer, comes from an overbalance of platonism in the "Christian" attitude. Western Christianity has created a false dichotomy between spirit and matter. Man is not *body* plus *soul* but soul incarnate. Otherworldliness, a preoccupation with the eternal at the expense of the temporal, a denial of esthetics because the arts are related to the physical, is an aberration of Christianity. Man's creation in the image of a transcendent God makes him no less a creature. This "creature-hood" was forever dramatized and established when God became flesh to live among us and to reveal His plan of salvation.

Pantheism can never be the answer, Dr. Schaeffer maintains. It makes man morally no better than the grass. It fails to explain the two faces of nature—her beneficence in the gentle rains and sunshine and her malevolence in the scorching desert or her storms and earthquakes. It tends to degrade rather than to elevate man. It gives no meaning to the individual.

What is required is a profoundly Christian perspective on nature and ecology. We need to reflect on the role of God and the role of man and on the role of Satan in the development of our present crisis. As we are overwhelmed with alarming environmental portents, we Christians need to remind ourselves of some basic biblical truths which can become anchors for our faith and guidelines for our acts and attitudes in the days ahead.

I would like to devote the rest of this chapter to a few of them.

God is the Creator. Modern science has taken away some of our awareness of this fundamental truth. The invention of the electronic microscope opened up an awesome new world of infinitesimal complexity as we discovered whole universes inside each living cell. All the intricacies of the DNA genetic chain and the microphysics of neutrons, negatons and protons have become routine. And as our scientists learn more and more about the secrets of life and matter, we seem conversely to become less and less aware of God's role in this marvelous process.

"In the beginning God created the heavens and the earth."

Most ecologists don't believe that. They are humanists who don't take God into account. To them the world is the product of an impersonal process of evolution. This attitude is apparent in the literature being produced on the subject. As the environmentalists study the depletion of the biosphere, the massive dislocation of its ecosystem, the runaway population and global food shortage, they are driven to near despair. "I have to say that I have been nearer to despair this year, 1968, than ever in my life," the then 63-year-old British author and statesman C.P. Snow declared.[4]

Some take refuge in utopianism.

The ecologists who authored, *Blueprint for Survival*, after analyzing in frightening detail all the factors that are contributing to the imminent collapse of our society and the exhaustion of our planet's resources, end their book with a chapter called, "A Legacy of Hope." They picture a stable society, semi-rural and decentralized, modest in its aspirations, altruistic in its distribution of wealth and resources, controlled in its growth and expansion, its population stabilized.

"Indeed, if we are capable of ensuring a relatively

smooth transition to it, we can be optimistic about providing our children with a way of life psychologically, intellectually, and esthetically more satisfying than the present one. And we can be confident that it will be sustainable, as ours cannot be, so that the legacy of despair we are about to leave them may at the last minute be changed to one of hope."[5]

Knowing human nature, to me this hope is slim!

The Christian need not evade grim reality, and it need not lead him to despair. Nor need he fall into utopian optimism. The fact is that the Christian faith is the only one which takes a realistic view of evil and bares it in all its ugliness. It offers man his surest hope. The Christian does not look for salvation within his own humanistic system. He does not depend upon the reformed altruism of his own fellow human beings. The Christian recognizes that God created the world in the first place, and He can recreate or renovate it when He so desires. The earth's hope is from outside itself. God is the Creator.

This does not exempt Christians from fulfilling their reponsibilities in the here and now. On the contrary, the Scriptures enjoin us to be faithful stewards, to make the best use of our opportunities, to share the Christ-life with our fellow men and to minister to their needs as best we can.

But if the night grows darker we need not despair. From the beginning God has foretold the destruction of this earth and the creation of a new one. "We have this promise," Peter assures us in 2 Peter 3:13, "and look forward to new heavens and a new earth, the home of justice."

He did it once. He can do it again. The Christian faith looks only to Him.

Man can be redeemed. The chauvinistic rape of nature by man is not a new thing. Its effects have become increasingly visible in the last century as population growth, economic affluence and accelerated technology have combined to deplete and contaminate the earth in proportions hitherto undreamt of. But the attitude is not new.

Nor is our crisis a product of Christian arrogance. It stems from attitudes which antedate the Christian era, even that of the kingdom of Israel. Man's arrogance towards nature began in the Garden of Eden when the Serpent whispered to the Woman that to disobey God and to taste the forbidden fruit would not bring death. "Of course you will not die," he lied. "God knows that as soon as you eat it, your eyes will be opened and you will be like gods knowing both good and evil" (Gen. 3:4,5).

When Adam and Eve considered themselves to be beyond the natural rules which God had laid down for their life, sin came into human experience. Man's hostility towards nature dates from that moment when Satan's lies convinced him that he could somehow be superior to the rest of creation and could exploit it for his own selfish ends.

Satan is still at large in the world. This concept was unpopular a generation ago. But today there seems to be a new consciousness of the demonic presence. Films like *The Exorcist* and *Rosemary's Baby* play on the feeling that demonism is real. The revival of Satanism and the popularity of the occult arts all point to a growing awareness of the forces of evil about us.

To the Christian this is nothing new. "Awake! be on the alert!" Peter long ago warned us. "Your enemy the devil, like a roaring lion, prowls round looking for someone to devour. Stand up to him, firm in faith . . . " (1 Pet. 5:8,9). From the times of ancient Job, God's people have held the belief that Satan is loose in the world. They believe he is seeking to thwart God's purposes and that this "ranging over the earth . . . from end to end" (Job 1:7) is mysteriously a part of God's plan for testing men and allowing them to choose voluntarily to love, obey and serve their Creator.

Judging from the present plight of the environment and the rather minimal awareness of it in Christian circles, it would seem fair to say that Satan is still busy blinding us to the facts, lying to us about the future, confusing our priorities, hiding from us the insidious flaws in our social systems, rationalizing away our sense of personal responsibility towards the environment and towards factors affecting it.

By pointing up the demonic factor, I do not want to minimize the human responsibility involved. Each man is responsible for his own sins and choices. But the demonic world so vividly brought to our attention by C.S. Lewis in *The Screwtape Letters* helps to explain why there are so few prophets among us. It clarifies why our theologians seem to be "majoring in the minors," and why many of us are constantly diverted into marginal debates on secondary issues. We have failed to perceive the mortality of the universe, the proximity of the judgment, the colossal ruin that is engulfing our fragile planet.

Man has made a mess of things. Each one of us is individually responsible before God for the moral perversity of our choices, our egocentric life-style, our complacency towards our starving fellows. Collectively we are under indictment for the abuse of our environment, the exhaustion of our resources, the contamination of our biosphere. Our stewardship has been a cataclysmic failure!

Is there yet hope? The Christian knows, "The conclusion of the matter is this: there is no condemnation for those who are united with Christ Jesus, because in Christ Jesus the life-giving law of the Spirit has set you free from the law of sin and death" (Rom. 8:1).

And the natural world shares this hope with us! "For the created universe waits with eager expectation for God's sons to be revealed. It was made the victim of frustration, not by its own choice, but because of him who made it so; yet always there was hope, because the universe itself is to be freed from the shackles of mortality and enter upon the liberty and splendor of the children of God. Up to the present, we know, the whole created universe groans in all its parts as if in the pangs of childbirth" (Rom. 8:19-22).

Sin is still rampant. Satan is not yet bound, although his days are numbered. In Jesus Christ there is salvation for mankind and hope for the created universe. By faith we can claim a relationship that will not release us from the experience of mortality, but which will assure us of forgiveness and renewal in the brightness of God's future.

Man and his environment can be redeemed!

God is the sovereign Lord of human history. History is the medium in which He has chosen to reveal Himself to us. Only in the context of sacred history is it possible to understand God. We know Him through His dealings with His people, the Israelites. It takes the sum of a nation's experience—its tragedies, victories, enslavement, liberation, perversity, endurance, rebellion and forgiveness, ethics, worship, laws and structures, prophecies and fulfillment, great saints and great sinners—to communicate something of who God is and what is His purpose for mankind.

Only in the context of history is it possible also to know Jesus Christ. He truly lived among men, exhibited the moral righteousness and power of the Holy Spirit, reacted to the pressures of contemporary events, died and rose again within the framework of human history. He revealed Himself as the Son of Man among men. He was what He claimed to be, the Mediator and Saviour of mankind.

God is the maker and unfolder of human history. "Blessed be God's name from age to age," sang the prophet Daniel, "for all wisdom and power are his. He changes seasons and times; he deposes kings and sets them up" (Dan. 2:20,21). He uses them for His purpose and then rewards or punishes them for their righteousness or evil doing.

Time after time in the experience of the Hebrew nation, the same great cycle is repeated. Israel falls into sin; God uses the pagan nations round about to punish His people for their disobedience; Israel repents; God liberates and restores His repentant people and punishes those who dealt cruelly with them. Examine the history of the Philistines, the Moabites, the Syrians, the Babylonians and the Egyptians. The fortunes of these neighboring nations interweave with those of Israel in great moral cycles which clearly demonstrate God's purpose in trying to prepare the Israelites for their coming Messiah.

In the Old Testament they were punished repeatedly for their idolatry and rejection of their sovereign God. This idolatrous tendency seems finally to have been extirpated by the Babylonian captivity. In the New Testament they are punished once again for rejecting the Messiah, Jesus Christ. They have been exiled throughout the world, enduring cruel persecution. There are hopeful signs of their eventual, full restoration. In God's time they will be both repentant and fully re-established as a nation, following which the Gentile nations, which have treated them so cruelly, will themselves be punished.

Even the reading of a secular book like *O Jerusalem!* gives one the inescapable sensation of being an observer of divine history. There is the passionate patriotism of both Jews and Arabs. Their bravery and bravura, their masterful diplomacy and banal quibblings, their precipitous recklessness and timid vacillation, their glorious victories and ignominious defeats, their astute strategies and their unbelievable blunders are dramatically recorded. All these factors make it seem as though once more He "who sitteth in the heavens" is deposing and setting up kings and rulers, controlling human history for His own ends, shaping things up to the final accomplishment of His eternal purpose.

As Christians we need again to steep ourselves in the sovereignty of God.

Jesus Christ is coming again soon. This truth is not very real to most of us. We believe it because it is stated unequivocally in the Bible. But its application to our own lives and to the priorities by which we are governed is rather vague.

Since I have been studying ecology, the second coming of Christ has taken on new meaning for me. I now see it as the first act in the coming drama of God's recreation of the new heavens and the new earth. The more I see of the ruin and depletion of our present planet, its pollution and asphyxiation, the more welcome is the prospect of Christ's soon return.

Scarcely any doctrine in the New Testament is more thoroughly attested to. Jesus promised His disciples that He would return bodily to the earth. Every one of the New Testament authors confirms this promise in his own inspired words. The Second Coming is presented as a comfort to those who are distressed and in sorrow (1 Thess. 4:18), as an encouragement to those who are impatient (Jas. 5:7), as an inducement to prayer and holy living (1 Pet. 4:7; 2 Pet. 3:11), and as a stimulus to faithful stewardship (2 Tim. 4:8) and the zealous preaching of the gospel (Matt. 24:14).

Some may ask, "Why has He delayed so long?" St. Peter anticipated this question. "In the last days there will come men who scoff at religion and live self-indulgent lives, and they will say: 'Where now is the promise of his coming? Our fathers have been laid to their rest, but still everything continues exactly as it has always been since the world began!' " (2 Pet. 3:3,4).

The answer is to be found in the analogy of Noah. His contemporaries scoffed and asked him the same question. Many came to regret their incredulity. According to some exegetes, the reason that Noah's grandfather Methuselah was the oldest man in history stems from God's mercy and reluctance to release upon the rebellious world the floods of punishment and cleansing. According to these exegetes, the name "Methuselah" can mean something to the effect that "when (he is) gone, (it) shall be," or shall come to pass. Thus the prolongation of Methuselah's life was a sign to Noah of God's constantly-extending mercy, giving men every chance to repent. The year that Methuselah died, the floods finally came. "It is not that the Lord is slow in fulfilling his promise, as some suppose, but that he is very patient with you, because it is not his will for any to be lost, but for all to come to repentance" (2 Pet. 3:9).

Every prolongation of this age of grace, every delay of Christ's return, provides more opportunity for the inhabitants of the earth to turn in repentance to God and seek His mercy. The repentance of Nineveh postponed God's judgment. A world-wide turning to Him today might be a part of His purpose in extending the age of grace.

The second coming of Christ is not just the first stage of liberation and renewal. It also announces the coming of judgment, and all who have refused His mercy will be punished in accord with their own free choice.

The ecological crisis highlights in a particular way these doctrines. The earth's biosphere did not just happen to evolve. It was created by the loving power of God. History is not just a fortuitous chain of events. God is the sovereign Lord of human history. He is working out His perfect purpose. Best of all, Jesus Christ is coming again soon to redeem His own and to prepare the world for judgment. This is a sure and blessed hope.

Footnotes

1. Lynn White, Jr., "The Historical Roots of Our Ecological Crisis," *Science* magazine, 1967.
2. Richard L. Means, "Why Worry About Nature?" in *Saturday Review*, December 2, 1967.
3. Francis Schaeffer, *Pollution and the Death of Man* (Wheaton: Tyndale, 1970). Page references are not given because we used the Spanish translation, *Polucion y la muerte del hombre, Enfoque cristiano a la ecologia* (El Paso: Casa Bautista, 1973).
4. Newspaper clipping, *New York Times*, November 13, 1968.
5. Edward Goldsmith et al., *Blueprint for Survival* (Boston: Houghton Mifflin Co., 1974), pp. 127, 128.

11

A CHRISTIAN LIFE-STYLE FOR TOMORROW

An increasing number of Christians are seriously concerned about how they should respond to the environmental crisis in terms of life-style.

There are environmentalists who return to the land, determined to learn for themselves how to restore the soil by organic farming and to wrest from it more nutritive crops. They want to isolate "natural" seeds, preparing for the day of cataclysm when our world-wide "miracle" hybrids are suddenly blighted by new pests and seeds become so costly as to be out of reach or simply unavailable. In a sense, they are training to be the survivors of cosmic disaster.

Other environmentalists make a fetish of fighting

one or more minor aspects of consumerism by refusing to purchase refined sugar or toothpaste in tubes. They may be right in their posture. But the battle of the environment is not likely to be won simply by the exercise of minor taboos or the survival of the fittest. A broader approach, with intelligent "trade-offs," is called for.

In this chapter we would like to suggest a series of principles to help guide Christians into a responsible personal posture *vis a vis* the crisis of our ecosphere. We are not prepared, nor are we competent, to suggest an environmental program for governments or political parties. We believe emphatically that they should have such programs. Our concern is to help fellow Christians, as individuals, grope towards a responsible participation in the stewardship of the creation God has placed at our disposal.

We choose to do this in the form of a decalogue for the ecological Christian.

Learn to Share

The Christian attitude must begin with love. Love one's neighbor. Show compassion for the suffering and deprived. We need a new searching of the Christian conscience to find out how we can express the love of Christ "unto the least of these my brethren" in these days when starvation threatens millions.

One family decided to skip a meal a week and to make the food available to an agency involved in feeding the destitute.

Another dedicated a second tithe to the starving, the first tithe going to the work of God in the Church.

Probably every family, whatever its economic

means, should at least do the equivalent of taking another hungry mouth into the family circle.

Barbara Ward suggests that "a certain percentage of after-tax income should go to works of justice at home and in the world, not as 'charity' but as another expression of justice . . . . And this examination of our own conscience will show us that in any case in environmental terms we must turn to a less selfish and consumptive ideal of life so that resources can be freed for the less fortunate and the pollutive risks of mass wealth diminished."[1]

Perhaps each of us should ask God, through His Holy Spirit, to motivate us to do our part and to reveal to us just how we should give expression to the love of Christ, reflected in us.

To us, the privileged, who by our careless exhaustion of the earth's diminishing resources have "robbed" the rest of the world of a chance for normal living, Paul's admonition to the Ephesian Christians is particularly apt: "The thief must give up stealing, and instead work hard and honestly with his own hands, so that he may have something to share with the needy" (Eph. 4:28).

Eat Intelligently

Considering that eating is the most ancient of man's skills or instincts, it is surprising how little we know about the science of nutrition! The last few years have produced a plethora of popular and valuable books on the subject. Many of them are well worth the reading.

From the ecological standpoint, there are a number of basic guidelines which every Christian housewife should keep in mind. Here are five suggestions:

Avoid processed foods as much as possible. Each step of processing reduces the nutritional value while at the same time it requires the consumption of energy and often helps deplete other reserves. Packaged potatoes have been dried and powdered, requiring large amounts of heat and energy. They are generally sold in foil bags or envelopes, using metal and the components of colored inks. They are less nourishing than plain, unprocessed spuds. By comparing prices per weight it is easy to see how much time and energy has gone into the processing and packaging.

Whole wheat bread is to be preferred over white bread and brown sugar over white sugar.

A second rule is to be careful of quantities. The waste of left-over food in our society is appalling. The average American home throws out 15 percent of its edible food. Most of us eat too much. This is merely a matter of habit. More careful discipline in determining the size of food servings will pay off in better health and greater economy. We should avoid excessive carbohydrates and non-foods like sugar, coffee and other stimulants. With so much of the world starving, how can we do less?

Eat low on the protein chain. Remember the Chinese pattern? Of 400 pounds of grain consumed per capita each year, 360 are eaten as grains and only 40 go to the cultivation of proteins in livestock and fowl. In the United States it's the other way around. About 85 percent of the 2,000 pounds of grain per capita goes to fatten animals.

There are two ways we can combat this waste. One

is to get as much of our protein as possible from lower on the food chain rather than from meat. Another is always to prefer leaner, pasture-fed meats over the "prime" cuts processed in wasteful feedlots.

If everyone in the United States were to reduce his meat diet by only one four-ounce hamburger of feedlot beef per week, this would save grain enough to feed half a million people for a year.

Favor home-grown products. Growing and canning or freezing your own fruits and vegetables can be a massive contribution to your own health as well as to the food coffers of the world. Care should be taken to provide the greatest possible variety in order to insure the body all the diverse minerals, vitamins, amino acids and other nutrients it needs.

Finally, dispose of kitchen wastes in ecologically sound ways. One of the best suggestions is multiple trash/garbage bins. Tin cans flattened out with their ends removed are acceptable for recycling. So are clear glass and green glass. Other bottles are returnable. But apart from the receptacles where these items are stored, each kitchen should have separate containers for trash (paper, string, etc.) and organic garbage. Ideally this latter could be composted for use as garden fertilizer.

Practice Frugality

For two generations Madison Avenue has tried to erase from our minds the ingrained thrift of our Yankee forebears. We have been bombarded with the advertisements of the "wastemakers," whose planned obsolescence requires us to buy and to keep buying and replacing. Propaganda whips up a sense of need where none existed, and stimulates acquisitiveness and impulse buying. We are victims of all the artificial expectations and aspirations of that distinctively North American economic invention known as consumerism.

Consumerism is simply a manifestation of materialism. It is basically self-centeredness. All of it is foreign to the truly Christian perspective. We need to pattern our life-style more after the Pennsylvania Mennonite than the Philadelphia banker. The depletion of resources around the world demands frugality as a way of life.

So what does that mean?

It means selecting wearability over style, in clothes and in automobiles. The next time you buy a car, choose the smallest and most durable and economical vehicle which will serve your purpose. Two cylinders and 500 pounds of steel should not be too much of a sacrifice, especially when you are saving yourself money, space and gasoline while you reduce pollution and conserve energy.

Frugality definitely means less gadgetry. Electric carving knives, pencil sharpeners, staplers—are these really necessary? Frugality means more attention to repairs and less to trade-ins. It means making things stretch and buying thoughtfully rather than impulsively. It means believing again that "a penny saved is a penny earned." And the slower the depletion of the world's nickel, tin, aluminum and petroleum (which also includes plastics), the longer our environmental cataclysm can be deferred.

Conserve Energy

It is easy to think that it makes little difference whether or not you leave another light bulb burning, or let the oven pre-heat while answering the phone, or use high-test rather than low-octane gasoline. In the aggregate these things count.

Quite probably the increasing costs of fuels will teach us all to be misers of energy. *Mankind at the Turning Point*, the Club of Rome's latest publication, maintains that the higher prices of oil are really a blessing in disguise and will slow down the depletion of reserves long enough to allow us to develop alternate energy sources. In Europe the high price of gasoline over the years has evolved a far more suitable standard of automobile size and performance than that of the United States.

A little imagination can produce a host of energy-saving patterns—car pools, less hot water in the tub, fewer night lights, fewer trips to the supermarket, carefully tuned engines and use of public transportation.

Abhor Pollution

"Don't be a litterbug!" People are more environment conscious today, although you wouldn't know it from the litter lining Miami's Biscayne Bay, or that left in the Dodger Stadium after a ball game. Our awareness of environmental pollution must be total and universal.

One of the most important ways of reducing pollution is to cut way back on the use of plastics, and to maximize the use and durability of whatever plastic items must be used. It is always preferable to use paper bags, returnable bottles, excelsior packing, paper picnic utensils, biodegradable paraphernalia and packaging.

Another important conservation practice is to use

soap instead of detergent. Soap and sunshine are much closer to nature's way of cleansing than are chemical detergents and a power-hungry dryer.

Crusade for ZPG (Zero Population Growth)

Without any doubt the basic aggravants of our environmental crisis are too many people and too much affluence (or technologically-produced affluence). Neither our ecological system nor our social structures nor our agricultural resources can stand the strain of our population doubling again every 20 or 30 years. This means that as Christians we must be intelligently informed on this most serious of contemporary problems and ready to do what we can to resolve it.

Many people think of the population problem as focusing on the underdeveloped countries of Latin America, Africa and Asia. And indeed it does. But the developed countries are not exempt from responsibility. Every new baby born in California represents at least twenty times more drain on the earth's resources than does a new baby in Colombia or Ceylon. It can therefore be argued logically that birth control is more necessary in the United States than in the Third World. Population control starts at home!

Some Christians have qualms about the morality of family planning. But, an educated ecological awareness quickly perceives the immorality of irresponsible procreation. While it is true that in the Bible no deliberate limitation of families was practiced, neither is it forbidden. The Scriptural teachings of responsibility in family relationships require an open mind towards fertility control. The physiological makeup of men and women elevates sexuality to far richer and higher significance than mere procreation. The Bible supports this position. The ways and means of family planning may be open to debate, but the need for it is self-evident.

It is our conviction that ecologically sophisticated Christians should go one step further and join the crusade for attaining a ZPG goal in every part of the globe. This calls for reading up on demography, wrestling with its problems, becoming a missionary of intelligent planning for family and society.

Encourage Decentralization

The British ecologists who authored *Blueprint for Survival* envision a stable society, rather than an expanding one, as the only viable alternative to ecological cataclysm. A stable society would be one characterized by (1) a minimum disruption of ecological processes; (2) maximum conservation of materials and energy; (3) a population in which recruitment equals loss, and (4) a social system in which the individual can enjoy, rather than feel restricted by, the first three conditions.

Essential to this type of society would be a decentralization of polity and economy at all levels. It means the formation of communities small enough to be reasonably self-regulating and self-supporting.

If the urban metropolis is the heaviest contributor to the disruption of our ecostystems, then the ecological crisis demands that we put the brakes on urban expansion and seek to decentralize the metropolis into more manageable social segments.

In personal terms, what does decentralization mean?

It means that whenever and wherever a choice is

possible, each Christian make decisions that would move him away from metropolitan to rural life-styles. He will raise his family where there are gardens and trees. He will have enough ground to be able to grow some flowers and vegetables. He will practice some organic gardening and some environmental conservation.

There will be more options along this line open to young people than to those already established in society with family and community responsibilities. But the principle is equally valid for all.

I am tempted here to quote our Lord's warning to His disciples: "So when you see 'the abomination of desolation' . . . then those who are in Judaea must take to the hills" (Matt. 24:15,16). Environmental understanding does not come easily to the urbanite who pounds the sidewalks of the city. It comes more readily to the one living in the world of nature. To one seeing, feeling, smelling, tasting the fruit of the land; observing the ravages of man, the denuding of mountains, the polluting of waters, the eroding of soils; learning to compost, to cultivate, to irrigate, to cooperate with the marvelous intricacies of God's ecosphere.

Farm Organically

Garden organically. Healthy plants, healthy animals and healthy people derive much of their health from healthy soil.

The "green revolution," with its programs of massive chemical fertilization and its intensive use of pesticides, is falling short of success in India and thousands are starving. Across the Himalayas in China, a nation beset by famine and starvation twenty-five years ago when its population numbered 400 million, is now well-fed and healthy with a margin of at least 10 percent above basic nutritional requirements for 800 million people. "Not only have cereals doubled (since 1949), but protein has increased twelve times. People now in China do not know what famine is. They do not know what starvation is, and I think that everybody is quite agreed that the Chinese have the most healthy population in the world."[2]

Sir Albert Howard calls Chinese truck-gardening "the most intensive agriculture which the world has so far seen . . . the classic example of a nation which has conserved the fertility of its soil."[3] For centuries the Chinese peasant has used all human and animal wastes, every leaf and twig, for developing organic compost to fortify the humus of his garden's soil.

A subscription to Robert Rodale's magazine, *Organic Gardening*,[4] will help one understand how to grow healthy roses or tomatoes and to cooperate with the organic processes of nature.

Plant Trees

As population grows, we need more oxygen. But the trees which are among its major producers, are diminishing. Every Sunday a small forest is consumed for one newspaper in New York City. "It is a very important and interesting newspaper," says Dr. Han Suyin, "and I read it religiously. But only eight percent of that paper goes to the news and 92 percent is devoted to advertising for the consumer society to spend more, to waste more and to buy more."[5]

Granted, some lumber and paper industries are trying to replace the wood used. But the conscientious companies make up a small fraction of the despoilers of our forests. Most of the world's forests have already been ravaged. We can do our bit today by planting trees. Even in a small lot, it is surprising how well certain trees fit. Planting them and helping them grow can be a very satisfying experience.

Be an Environmental Citizen

A growing ecological awareness has already produced in the United States several decisions with environmental implications which otherwise would have gone differently. The nuclear test ban treaty was negotiated. The supersonic transport project was scratched. The Army was deterred from dumping containers of toxic chemicals into the Atlantic and from transporting nerve gas across the country. Other victories have been won in Oregon, Florida, Alaska, Hawaii and California. They may be small victories. But they are symptomatic of a growing awareness on the part of the public that there is no such thing as a free lunch in ecology.

Communicating to one's congressman and senators is a minimal Christian responsibility. In order to act responsibly some reading is necessary. For this purpose we would suggest Barry Commoner's, *The Closing Circle* (Bantam, 1971/1974), as an excellent and very readable general source. The environmental sections of several magazines are also helpful. And there is a flood of valuable material available from the United Nations, especially from UNESCO, FAO and the Center for Economic and Social Information/OPI.

As has already been noted, industry is one of the environment's most ruthless despoilers and the major depleter of its mineral and energy reserves. Much of industry's touted progress and profit is realized at the expense of our environment. Somehow industry must be made to figure environmental factors into its cost accounting. Through ecological loss and the cost of environmental repair, we are in effect subsidizing industry. If environmental cost is not computed in, its total profits may be illusory.

As an ecologically oriented Christian, it is my duty to be informed in this area when it relates to a particular problem in my own neighborhood. I can encourage the application of public pressure in favor of conservation and the preservation of our common resources.

Assistance can come from the Scientists' Institute for Public Information (SIPI—30 East 68th Street, New York, New York 10021). SIPI publishes *Environment* magazine, an important educational asset.

In summary, we have no alternative but to live in ways that are more ecological and more Christian: learning "to conserve rather than consume, to abstain rather than indulge, to share rather than to hoard, to realize that the welfare of others is indistinguishable from our own."[6]

Each individual Christian will have to work out his own life-style and determine his own ecological priorities. He should not let the fact that he can't do *much* deter him from doing *something.*

Footnotes

1. Barbara Ward, "Environment, Population and Development" in WACC *Journal*, Frankfurt/Main, Germany, January—March, 1974.
2. Han Suyin, "Controlling the population . . . is not the cure" in *Development Forum*, Centre for Economic and Social Information/OPI, Geneva, September/October, 1974, p. 5.
3. Sir Albert Howard, *The Soil and Health* (New York: Schocken, 1947-1972), p. 39.
4. *Organic Gardening*, 33 E. Minor Street, Emmaus, Penna. 18049.
5. Han Suyin, op. cit., p. 5.
6. Paul Ehrlich interview, *Project Survival* (Chicago: Playboy Press, 1971), p. 104.

12

CHRISTIAN PRIORITIES FOR THE "LAST DAYS"

I'm only one person—what can I do?

The environmental crisis which the world is now experiencing is so serious, so profound, so complex, that it is easy to build a case for doing nothing about it.

"After all," you say, "I am only one small person. Nothing I do is likely to be that important."

"That's right," we would reply. "By yourself there's not much you can do to hold back galloping population or massive pollution. But you can be aware of what's happening. That's all we would ask you to do for the moment. If you are aware of the causes and effects of environmental collapse, you'll find that your whole attitude will change. Your priorities will shift and your life will become in its own way an effective force for good in the world."

It's something like an awareness of your own mortality.

A couple of years ago, when I was going through

a period of particularly hard work and stress, I experienced a temporary memory blackout. Frankly, it scared me. As it turned out, there was nothing seriously wrong. Any heart or blood-pressure problem can remind you, as the memory blackout hammered home to me, that you are mortal. As a consequence, for two years now I have been living with a new awareness of my own mortality, an awareness that cannot but affect my priorities, my decisions related to family, career and religious involvements.

Most of us are at least intellectually conscious of our own mortality. We also realize that civilizations are mortal. We have studied the rise and fall of cultures and empires. We recognize that our own occidental culture must ultimately lose momentum and grind to a halt, or to a lesser role in human history.

It is the mortality of Planet Earth that we want to emphasize here.

Our terrestrial spaceship is finite, not infinite. It is temporal, not eternal. We have already seen how near it is to overloading, depletion and collapse. All we ask is that as Christians we develop an awareness of its mortality.

Our decisions and our life-style will be directed towards restoring and preserving rather than destroying our environment. Our activities will be oriented towards improving our social structures, distributing our resources more equitably, seeking justice for our fellows, serving them as we are able.

"But there's so little *I* can do," you repeat.

It may not seem like much to you, but you can point your life in the right direction by obeying your Lord. "Help one another to carry these heavy loads, and in this way you will fulfill the law of Christ" (Gal.

6:2). Even in small things your obedience will be an act of worship to God and a testimony to others.

A new awareness of the earth's mortality will make us realize how short the time is. It forces us to re-examine our own goals and objectives to be sure that we are "redeeming the time" for our Lord and Saviour, making the best use of it we can.

But if the crisis is hastening Christ's return, shouldn't I just let it get worse?

There are those who see famines, wars and disasters as being the fulfillment of biblical prophecy. These things have been prophesied, they say, and they cannot be avoided. We should simply rejoice that the Lord's return is drawing near.

Senator Mark Hatfield, one of the most knowledgeable of Christian environmentalists today, objects strongly to this line of reasoning.

"I do not want to get into a discussion about eschatology, and all the various doctrines about the last days which have been such a source of division among us," he has said. "But let us be agreed about one central biblical truth. We are never told to sit by and watch the world destroy itself in its inhumanity and sin, and console ourselves with the prediction that the end of all things must be just around the corner."[1]

Theologian Vernon Grounds put it this way: "We may believe that history will end in utter destruction before the New Jerusalem comes into being. But that should not deter us from ministering to the world's suffering and need any more than the knowledge of the eventual death of every person would lead us to abandon any ministry to sickness and disease."[2]

As a matter of fact, in the middle of His discourse

on the end times, Jesus Himself made it clear that we will be judged severely on whether or not we turn our backs on the needy of the world:

"When the Son of Man comes in his glory and all the angels with him, he will sit in state on his throne," Jesus prophesied, "with all the nations gathered before him. He will separate men into two groups, as a shepherd separates the sheep from the goats, and he will place the sheep on his right hand and the goats on his left. Then the king will say to those on his right hand, 'You have my Father's blessing; come, enter and possess the kingdom that has been ready for you since the world was made. For when I was hungry, you gave me food; when thirsty, you gave me drink . . . . I tell you this: anything you did for one of my brothers here, however humble, you did for me' " (Matt. 25:31-40).

Hatfield has little patience for those who rationalize their inactivity and unwillingness to help the needy on the basis of eschatological prophecy. "Let no more be heard about people being poor or suffering because it is 'God's will,' and thus there is nothing we should do," he chides and quotes a pertinent comment of Thomas Merton:

"It is easy enough to tell the poor to accept their poverty as God's will when you yourself have warm clothes and plenty of food and medical care and a roof over your head and no worry about the rent. But if you want them to believe you—try to share some of their poverty and see if you can accept it as God's will yourself!"[3]

Is there any hope for the earth?

Some years ago the German theologian Edmund Schlink, of Heidelberg University, shocked an ecumenical conference which had gathered to consider "Jesus Christ, the Hope of the World."

"If in our thinking about this subject we place the emphasis on the preservation of this threatened world, we shall miss the point. If we expect Christ to ensure this world so that men may continue undisturbed in their pursuit of liberty, may carry on their business and seek an improvement in their standards of living, then Christ is not the hope of the world, but the end of all the world's hopes." The coming of Christ as the hope of the world means also the end of the world as we know it.[4]

A careful look at the earth does not inspire hope. We are running out of fuel, food, water, air and minerals—just about everything but people. "We are the most knowledgeable generation in the history of the world," affirms Dr. Samuel Moffett, but all we seem to be able to do with our vaunted technology is to "build another Tower of Babel, booby-trapped with nuclear weapons. We have wasted the good earth the Lord has given us, polluted his clean air, fouled the streams and brooks. Our cities are a stink and a disgrace. Look at the world as it really is," continues Moffett, "and if you look only at the world, don't babble about hope." "The earth shall perish," says the Old Testament (Ps. 102:26). It will be "burned up," adds the New Testament (2 Pet. 3:10).

Is there, then, any hope for the earth? The answer is "yes," but it is the radical hope that is in a seed—it must die before it can burst into a springtime of new life. And it does not come from human sources.

"One of the most important lessons in the Bible," states Moffett, "is that hope is not confined to any one point in space or time. It is tied to a person: 'Jesus

Christ is the same yesterday, today, and forever' " (Heb. 13:8).

"Things do look bad these days. But if you have given up hope because *today* is so bad, look back about 1900 years to the darkest day the world has known: the day man took the hope of the world and stripped him, beat him, and crucified him. The dead shuddered, hell broke loose, and for one agonizing instant, a moment never to be repeated, the whole human race was utterly lost. 'My God, my God, why have you forsaken me?' came a cry from the cross, from the second Adam.

"But God took that most hopeless of all days," Moffett goes on, "and made it the hinge of history, not its end. Man's *curse* is that without God he takes each new shining hope and turns it into an engine of his own destruction. Man's *hope* lies in the fact that God does just the opposite."[5]

The world's basic trouble today is not that it is running out of physical resources, but that it is running out of hope. And it is running out of hope because it is placing its hope in the wrong thing—in conserving and preserving its physical resources. If the ecological crisis breeds despair in our souls, the hope of salvation cannot come from a utopian dream of reversing the pattern. It can come only from the sovereign Lord of all, Jesus Christ the Hope of the world.

Some Christians, however, manage to miss one of the key points about hope, namely, that it carries with it a mission. We are not offering in Christ's name a mere "pie in the sky bye-and-bye." The gospel cannot be made to say: "Since there is no hope for the world, the sooner you are out of it the better. Die, therefore, and receive the reward laid up for you in heaven." Such a "gospel" deserves to be met with scorn, not hope.

"Jesus is the hope of the world not simply because he calls us to future glory," Moffett points out. "He is the hope of the world because he also laid aside his glory to share the hungers of the hungry and feed them, to suffer the weakness of the sick and heal them, to take on himself the injustices of the oppressed and overcome them.

"In this day of increasingly serious shortages," Moffett reasons, "it is time for Christians to recognize that any witness which has nothing to say about the consuming hungers of two-thirds of the world's peoples is a witness neither inspired by Christ who fed the multitudes, nor one that will win the hearts of the multitudes he died to save. When people are starving, they look for bread, not for preaching."

"But let's not distort the gospel the other way, either," he warns. "Our hope is the hope of salvation, centered in the resurrection of Jesus Christ. For the Church to settle for any lesser hope, whether by technological advance or by social action, is a betrayal of the faith and no ultimate service to mankind. Finding enough food, water and oil to keep this world going, and saying that's enough, is like throwing a life preserver to a man who has fallen overboard from an ocean liner and not stopping to pick him up. It will keep him from drowning, perhaps, only to doom him to the wind, the sun and the sharks. The life preserver keeps him alive to be rescued, but what finally counts is the rescue.

"So with our mission. When Christ bids us give

water to the thirsty, he adds, give it 'in my name.' For there is a deeper thirst than the physical—a thirst that only Christ can satisfy."[6]

There is no other hope for the earth and its inhabitants.

In view of these things, what should be my most urgent concern?

In his second letter, St. Peter asks and answers the same question: "Since the whole universe is to break up in this way, think what sort of people you ought to be, what devout and dedicated lives you should live!" (2 Pet. 3:11).

"Devout" and "dedicated." This says something about the Godward and manward dimensions of our religion and service. Devotion to God embraces obedience, worship, dependence; while dedication to God's objectives and to our fellow men means nothing short of giving ourselves completely, selflessly, on their behalf.

I would want to be sure that my relationship to God was a real, life-giving, power-full and sustaining one. Such a relationship is possible only through faith in God's Son, Jesus Christ. "For there is one God, and also one mediator between God and men, Christ Jesus, himself man, who sacrificed himself to win freedom for all mankind, so providing, at the fitting time, proof of the divine purpose" (1 Tim. 2:5,6).

Being sure of my saving relationship to God, I would try to learn all I could about God's standards for the Christian life. I would try to live up to them. Love is the first of the Christian virtues, but the fruit of the Spirit includes many more characteristics: joy, peace, patience, kindness, goodness, fidelity, gentleness, self-control. In this connection Peter says, "The end of all things is upon us, so you must lead an ordered and sober life, given to prayer. Above all, keep your love for one another at full strength, because love cancels innumerable sins" (1 Pet. 4:7,8).

We should all be concerned with getting out the message of God's revelation to the world in Jesus Christ. We are here to bear witness to the truth, to fulfill our Lord's great and last commission, to see that the message of Christ's gospel covers the earth like the waters cover the sea.

Jesus' teaching was replete with allusions to man's role in this life as one of stewardship. The parables of the talents, the pounds, the vineyards, the two sons, the foolish virgins and the pearl of great price all reflect on the quality of Christian stewardship.

I don't know what particular talents God may have entrusted to you. But I do know that He expects you to use them for His glory. We each play a different instrument. It is important that we each follow the score, so that the full orchestration may be harmonious, powerful, convincing, reaching an inner chord of response from all men everywhere. Serving, teaching, prophesying, healing, administering, reconciling, interpreting, preaching, God expects us all to respond to the Holy Spirit's direction and to give ourselves in devotion and dedicated service to Him.

We have been entrusted with the Gospel of grace, the good news of Jesus Christ's mediation on our behalf, the message of pardon, new life and salvation. We have been clearly commanded to share this treasure with those who do not know Christ as their personal Saviour.

Your priorities will be affected by your responsibilities and the particular burdens which God may have placed on your heart. If you are a government official your response to the ecological crisis will be different from that of a farmer or the pastor of a church. But all of us share the stewardship of the gospel. We must get this message out!

"For us as individuals," concludes Senator Hatfield, "stewardship means molding our own lives to the shape of the Man Crucified. Christ beckons us to far more than simply charitable giving. He calls us to love, and to give in a way that changes the shape of our lives. That love led Him to the cross. And it will lead us to the cross if we follow Him, to pour ourselves out for the sake of others."[7]

These words of 1 John should speak to our deepest selves and send us on our mission:

"We for our part have crossed over from death to life; this we know, because we love our brothers. The man who does not love is still in the realm of death .... It is by this that we know what love is: that Christ laid down his life for us. And we in turn are bound to lay down our lives for our brothers. But if a man has enough to live on, and yet when he sees his brother in need shuts up his heart against him, how can it be said that the divine love dwells in him?" (3:14-17).

If we mold our lives to His, we shall look forward with joy to His coming and to the new heavens and the new earth "wherein justice shall reign."

" 'Yes, I am coming soon,' Jesus has promised, 'and bringing my recompense with me, to requite everyone according to his deeds! I am the Alpha and the Omega, the first and the last, the beginning and the end .... '

" 'Come!' say the Spirit and the bride.

" 'Come!' let each hearer reply.

"Come forward, you who are thirsty; accept the water of life, a free gift to all who desire it . . . .

"He who gives this testimony speaks: 'Yes, I am coming soon!'

"Amen. Come, Lord Jesus!

"The grace of the Lord Jesus be with you all" (Rev. 22:12,13,17,20,21).

Footnotes

1. Mark Hatfield, "Thanksgiving 1974: Feast or Famine?" in *Eternity* magazine, November, 1974.
2. Quoted by Mark Hatfield, op. cit.
3. Quoted by Mark Hatfield, op. cit.
4. Quoted by Samuel Moffett in "Where in the World Is Hope?," address delivered at Inter-Varsity Urbana conference, December, 1973, condensed in *The Christian Reader*, November/December, 1974.
5. Samuel Moffett, op. cit.
6. *Ibid.*
7. Mark Hatfield, op. cit.

INDEX